Science and Method

Science and Method

HENRI POINCARÉ
FRANCIS MAITLAND, TRANSLATOR

COSIMO CLASSICS
NEW YORK

Science and Method

Science and Method was originally published in 1914.

For information, address:
P.O. Box 416, Old Chelsea Station
New York, NY 10011

or visit our website at:
www.cosimobooks.com

Ordering Information:
Cosimo publications are available at online bookstores. They may also be purchased for educational, business or promotional use:
- *Bulk orders:* special discounts are available on bulk orders for reading groups, organizations, businesses, and others. For details contact Cosimo Special Sales at the address above or at info@cosimobooks.com.
- *Custom-label orders:* we can prepare selected books with your cover or logo of choice. For more information, please contact Cosimo at info@cosimobooks.com.

Cover Design by www.popshopstudio.com

ISBN: 978-1-61640-254-9

If the scientist had an infinity of time at his disposal, it would be sufficient to say to him, "Look, and look carefully." But, since he has not time to look at everything, and above all to look carefully, and since it is better not to look at all than to look carelessly, he is forced to make a selection.

—from *Science and Method*

CONTENTS

INTRODUCTION.

In this work I have collected various studies which are more or less directly concerned with scientific methodology. The scientific method consists in observation and experiment. If the scientist had an infinity of time at his disposal, it would be sufficient to say to him, "Look, and look carefully." But, since he has not time to look at everything, and above all to look carefully, and since it is better not to look at all than to look carelessly, he is forced to make a selection. The first question, then, is to know how to make this selection. This question confronts the physicist as well as the historian; it also confronts the mathematician, and the principles which should guide them all are not very dissimilar. The scientist conforms to them instinctively, and by reflecting on these principles one can foresee the possible future of mathematics.

We shall understand this still better if we observe the scientist at work; and, to begin with, we must have some acquaintance with the psychological mechanism of discovery, more especially that of mathematical discovery. Observation of the mathematician's method of working is specially instructive for the psychologist.

In all sciences depending on observation, we must

reckon with errors due to imperfections of our senses and of our instruments. Happily we may admit that, under certain conditions, there is a partial compensation of these errors, so that they disappear in averages. This compensation is due to chance. But what is chance? It is a notion which is difficult of justification, and even of definition; and yet what I have just said with regard to errors of observation, shows that the scientist cannot get on without it. It is necessary, therefore, to give as accurate a definition as possible of this notion, at once so indispensable and so elusive.

These are generalities which apply in the main to all sciences. For instance, there is no appreciable difference between the mechanism of mathematical discovery and the mechanism of discovery in general. Further on I approach questions more particularly concerned with certain special sciences, beginning with pure mathematics.

In the chapters devoted to them, I am obliged to treat of somewhat more abstract subjects, and, to begin with, I have to speak of the notion of space. Every one knows that space is relative, or rather every one says so, but how many people think still as if they considered it absolute. Nevertheless, a little reflection will show to what contradictions they are exposed.

Questions concerning methods of instruction are of importance, firstly, on their own account, and secondly, because one cannot reflect on the best method of imbuing virgin brains with new notions without, at the same time, reflecting on the manner in which these notions have been acquired by our ancestors, and consequently on their true origin—that is, in reality, on their true nature. Why is it that, in most cases, the definitions which satisfy scientists mean nothing at all to children? Why is it necessary to give them other definitions? This is the question I have set myself in the chapter which follows, and its solution might, I think, suggest useful reflections to philosophers interested in the logic of sciences.

On the other hand, there are many geometricians who believe that mathematics can be reduced to the rules of formal logic. Untold efforts have been made in this direction. To attain their object they have not hesitated, for instance, to reverse the historical order of the genesis of our conceptions, and have endeavoured to explain the finite by the infinite. I think I have succeeded in showing, for all who approach the problem with an open mind, that there is in this a deceptive illusion. I trust the reader will understand the importance of the question, and will pardon the aridity of the pages I have been constrained to devote to it.

The last chapters, relating to mechanics and astronomy, will be found easier reading.

Mechanics seem to be on the point of undergoing a complete revolution. The ideas which seemed most firmly established are being shattered by daring innovators. It would certainly be premature to decide in their favour from the start, solely because they are innovators; but it is interesting to state their views, and this is what I have tried to do. As far as possible I have followed the historical order, for the new ideas would appear too surprising if we did not see the manner in which they had come into existence.

Astronomy offers us magnificent spectacles, and raises tremendous problems. We cannot dream of applying the experimental method to them directly; our laboratories are too small. But analogy with the phenomena which these laboratories enable us to reach may nevertheless serve as a guide to the astronomer. The Milky Way, for instance, is an assemblage of suns whose motions appear at first sight capricious. But may not this assemblage be compared with that of the molecules of a gas whose properties we have learnt from the kinetic theory of gases? Thus the method of the physicist may come to the aid of the astronomer by a side-track.

Lastly, I have attempted to sketch in a few lines the history of the development of French geodesy. I have shown at what cost, and by what persevering efforts and often dangers, geodesists have secured for us the few notions we possess about the shape of the earth. Is this really a question of method? Yes, for this history certainly teaches us what precautions must surround any serious scientific operation, and what time and trouble are involved in the conquest of a single new decimal.

BOOK I.

THE SCIENTIST AND SCIENCE.

I.

THE SELECTION OF FACTS.

TOLSTOI explains somewhere in his writings why, in his opinion, "Science for Science's sake" is an absurd conception. We cannot know all the facts, since they are practically infinite in number. We must make a selection; and that being so, can this selection be governed by the mere caprice of our curiosity? Is it not better to be guided by utility, by our practical, and more especially our moral, necessities? Have we not some better occupation than counting the number of lady-birds in existence on this planet?

It is clear that for him the word utility has not the meaning assigned to it by business men, and, after them, by the greater number of our contemporaries. He cares but little for the industrial applications of science, for the marvels of electricity or of automobilism, which he regards rather as hindrances to moral progress. For him the useful is exclusively what is capable of making men better.

It is hardly necessary for me to state that, for my part, I could not be satisfied with either of these ideals. I have no liking either for a greedy and narrow plutocracy, or for a virtuous unaspiring democracy, solely occupied in turning the other cheek, in which we should find good people devoid of curiosity, who, avoiding all excesses, would not die of any disease—save boredom. But it is all a matter of taste, and that is not the point I wish to discuss.

None the less the question remains, and it claims our attention. If our selection is only determined by caprice or by immediate necessity, there can be no science for science's sake, and consequently no science. Is this true? There is no disputing the fact that a selection must be made: however great our activity, facts outstrip us, and we can never overtake them; while the scientist is discovering one fact, millions and millions are produced in every cubic inch of his body. Trying to make science contain nature is like trying to make the part contain the whole.

But scientists believe that there is a hierarchy of facts, and that a judicious selection can be made. They are right, for otherwise there would be no science, and science does exist. One has only to open one's eyes to see that the triumphs of industry, which have enriched so many practical men, would never have seen the light if only these practical men had existed, and if they had not been preceded by disinterested fools who died poor, who never thought of the useful, and yet had a guide that was not their own caprice.

What these fools did, as Mach has said, was to save their successors the trouble of thinking. If they had worked solely in view of an immediate application, they would have left nothing behind them, and in face of a new requirement, all would have had to be done again. Now the majority of men do not like thinking, and this is perhaps a good thing, since instinct guides them, and very often better than reason would guide a pure intelligence, at least whenever they are pursuing an end that is immediate and always the same. But instinct is routine, and if it were not fertilized by thought, it would advance no further with man than with the bee or the ant. It is necessary, therefore, to think for those who do not like thinking, and as they are many, each one of our thoughts must be useful in as many circumstances as possible. For this reason, the more general a law is, the greater is its value.

This shows us how our selection should be made. The most interesting facts are those which can be used several times, those which have a chance of recurring. We have been fortunate enough to be born in a world where there are such facts. Suppose that instead of eighty chemical elements we had eighty millions, and that they were not some common and others rare, but uniformly distributed. Then each time we picked up a new pebble there would be a strong probability that it was composed of some unknown substance. Nothing that we knew of other pebbles would tell us anything about it. Before each new object we should be like a new-born child; like him we could but obey our caprices or our necessities. In such a world there would be no science, perhaps thought and even life would be impossible, since evolution could not have developed the instincts of self-preservation. Providentially it is not so; but this blessing, like all those to which we are accustomed, is not appreciated at its true value. The biologist would be equally embarrassed if there were only individuals and no species, and if heredity did not make children resemble their parents.

Which, then, are the facts that have a chance of recurring? In the first place, simple facts. It is evident that in a complex fact many circumstances are united by chance, and that only a still more improbable chance could ever so unite them again. But are there such things as simple facts? and if there are, how are we to recognize them? Who can tell that what we believe to be simple does not conceal an alarming complexity? All that we can say is that we must prefer facts which appear simple, to those in which our rude vision detects dissimilar elements. Then only two alternatives are possible; either this simplicity is real, or else the elements are so intimately mingled that they do not admit of being distinguished. In the first case we have a chance of meeting the same simple fact again, either in all its purity, or itself entering as an element into some complex whole. In the second case the intimate mixture has similarly a greater chance of being reproduced than a heterogeneous assemblage. Chance can mingle, but it cannot unmingle, and a combination of various elements in a well-ordered edifice in which something can be distinguished, can only be made deliberately. There is, therefore, but little chance that an assemblage in which different things can be distinguished should ever be reproduced. On the other hand, there is great probability that a mixture which appears homogeneous at first sight will be reproduced several times. Accordingly facts which appear simple, even if they are not so in reality, will be more easily brought about again by chance.

It is this that justifies the method instinctively adopted by scientists, and what perhaps justifies it still better is that facts which occur frequently appear to us simple just because we are accustomed to them.

But where is the simple fact? Scientists have tried to find it in the two extremes, in the infinitely great and in the infinitely small. The astronomer has found it because the distances of the stars are immense, so great that each of them appears only as a point and qualitative differences disappear, and because a point is simpler than a body which has shape and qualities. The physicist, on the other hand, has sought the elementary phenomenon in an imaginary division of bodies into infinitely small atoms, because the conditions of the problem, which undergo slow and continuous variations as we pass from one point of the body to another, may be regarded as constant within each of these little atoms. Similarly the biologist has been led instinctively to regard the cell as more interesting than the whole animal, and the event has proved him right, since cells belonging to the most diverse organisms have greater resemblances, for those who can recognize them, than the organisms themselves. The sociologist is in a more embarrassing position. The elements, which for him are men, are too dissimilar, too variable, too capricious, in a word, too complex themselves. Furthermore, history does not repeat itself; how, then, is he to select the interesting fact, the fact which is repeated? Method is precisely the selection of facts, and accordingly our first care must be to devise a method. Many have been devised because none holds the field undisputed. Nearly every sociological thesis proposes a new method, which, however, its author is very careful not to apply, so that sociology is the science with the greatest number of methods and the least results.

It is with regular facts, therefore, that we ought to begin; but as soon as the rule is well established, as soon as it is no longer in doubt, the facts which are in complete conformity with it lose their interest, since they can teach us nothing new. Then it is the exception which becomes important. We cease to look for resemblances, and apply ourselves before all else to differences, and of these differences we select first those that are most accentuated, not only because they are the most striking, but because they will be the most instructive. This will be best explained by a simple example. Suppose we are seeking to determine a curve by observing some of the points on it. The practical man who looked only to immediate utility would merely observe the points he required for some special object; these points would be badly distributed on the curve, they would be crowded together in certain parts and scarce in others, so that it would be impossible to connect them by a continuous line, and they would be useless for any other application. The scientist would proceed in a different manner. Since he wishes to study the curve for itself, he will distribute the points to be observed regularly, and as soon as he knows some of them, he will join them by a regular line, and he will then have the complete curve. But how is he to accomplish this? If he has determined one extreme point on the curve, he will not remain close to this extremity, but will move to the other end. After the two extremities, the central point is the most instructive, and so on.

Thus when a rule has been established, we have first

to look for the cases in which the rule stands the best chance of being found in fault. This is one of many reasons for the interest of astronomical facts and of geological ages. By making long excursions in space or in time, we may find our ordinary rules completely upset, and these great upsettings will give us a clearer view and better comprehension of such small changes as may occur nearer us, in the small corner of the world in which we are called to live and move. We shall know this corner better for the journey we have taken into distant lands where we had no concern.

But what we must aim at is not so much to ascertain resemblances and differences, as to discover similarities hidden under apparent discrepancies. The individual rules appear at first discordant, but on looking closer we can generally detect a resemblance; though differing in matter, they approximate in form and in the order of their parts. When we examine them from this point of view, we shall see them widen and tend to embrace everything. This is what gives a value to certain facts that come to complete a whole, and show that it is the faithful image of other known wholes.

I cannot dwell further on this point, but these few words will suffice to show that the scientist does not make a random selection of the facts to be observed. He does not count lady-birds, as Tolstoi says, because the number of these insects, interesting as they are, is subject to capricious variations. He tries to condense a great deal of experience and a great deal of thought into a small volume, and that is why a little book on physics contains so many past experiments, and a thousand times as many possible ones, whose results are known in advance.

But so far we have only considered one side of the question. The scientist does not study nature because it is useful to do so. He studies it because he takes pleasure in it, and he takes pleasure in it because it is beautiful. If nature were not beautiful it would not be worth knowing, and life would not be worth living. I am not speaking, of course, of that beauty which strikes the senses, of the beauty of qualities and appearances. I am far from despising this, but it has nothing to do with science. What I mean is that more intimate beauty which comes from the harmonious order of its parts, and which a pure intelligence can grasp. It is this that gives a body a skeleton, so to speak, to the shimmering visions that flatter our senses, and without this support the beauty of these fleeting dreams would be imperfect, because it would be indefinite and ever elusive. Intellectual beauty, on the contrary, is self-sufficing, and it is for it, more perhaps than for the future good of humanity, that the scientist condemns himself to long and painful labours.

It is, then, the search for this special beauty, the sense of the harmony of the world, that makes us select the facts best suited to contribute to this harmony; just as the artist selects those features of his sitter which complete the portrait and give it character and life. And there is no fear that this instinctive and unacknowledged preoccupation will divert the scientist from the search for truth. We may dream of a harmonious world, but how far it will fall short of the real world! The Greeks, the greatest artists that ever were, constructed a heaven for themselves; how poor a thing it is beside the heaven as we know it!

It is because simplicity and vastness are both beautiful that we seek by preference simple facts and vast facts; that we take delight, now in following the giant courses of the stars, now in scrutinizing with a microscope that prodigious smallness which is also a vastness, and now in seeking in geological ages the traces of a past that attracts us because of its remoteness.

Thus we see that care for the beautiful leads us to the same selection as care for the useful. Similarly economy of thought, that economy of effort which, according to Mach, is the constant tendency of science, is a source of beauty as well as a practical advantage. The buildings we admire are those in which the architect has succeeded in proportioning the means to the end, in which the columns seem to carry the burdens imposed on them lightly and without effort, like the graceful caryatids of the Erechtheum.

Whence comes this concordance? Is it merely that things which seem to us beautiful are those which are best adapted to our intelligence, and that consequently they are at the same time the tools that intelligence knows best how to handle? Or is it due rather to evolution and natural selection? Have the peoples whose ideal conformed best to their own interests, properly understood, exterminated the others and taken their place? One and all pursued their ideal without considering the consequences, but while this pursuit led some to their destruction, it gave empire to others. We are tempted to believe this, for if the Greeks triumphed over the barbarians, and if Europe, heir of the thought of the Greeks, dominates the world, it is due to the fact that the savages loved garish colours and the blatant noise of the drum, which appealed to their senses, while the Greeks loved the intellectual beauty hidden behind sensible beauty, and that it is this beauty which gives certainty and strength to the intelligence.

No doubt Tolstoi would be horrified at such a triumph, and he would refuse to admit that it could be truly useful. But this disinterested pursuit of truth for its own beauty is also wholesome, and can make men better. I know very well there are disappointments, that the thinker does not always find the serenity he should, and even that some scientists have thoroughly bad tempers.

Must we therefore say that science should be abandoned, and morality alone be studied? Does any one suppose that moralists themselves are entirely above reproach when they have come down from the pulpit?

II.

THE FUTURE OF MATHEMATICS.

If we wish to foresee the future of mathematics, our proper course is to study the history and present condition of the science.

For us mathematicians, is not this procedure to some extent professional? We are accustomed to *extrapolation*, which is a method of deducing the future from the past and the present; and since we are well aware of its limitations, we run no risk of deluding ourselves as to the scope of the results it gives us.

In the past there have been prophets of ill. They took pleasure in repeating that all problems susceptible of being solved had already been solved, and that after them there would be nothing left but gleanings. Happily we are reassured by the example of the past. Many times already men have thought that they had solved all the problems, or at least that they had made an inventory of all that admit of solution. And then the meaning of the word solution has been extended; the insoluble problems have become the most interesting of all, and other problems hitherto undreamed of have presented themselves. For the Greeks a good solution was one that employed only rule and compass; later it became one obtained by the extraction of radicals, then one in which algebraical functions and radicals alone figured. Thus the pessimists found themselves continually passed over, continually forced to retreat, so that at present I verily believe there are none left.

My intention, therefore, is not to refute them, since they are dead. We know very well that mathematics will continue to develop, but we have to find out in what direction. I shall be told "in all directions," and that is partly true; but if it were altogether true, it would become somewhat alarming. Our riches would soon become embarrassing, and their accumulation would soon produce a mass just as impenetrable as the unknown truth was to the ignorant.

The historian and the physicist himself must make a selection of facts. The scientist's brain, which is only a corner of the universe, will never be able to contain the whole universe; whence it follows that, of the innumerable facts offered by nature, we shall leave some aside and retain others. The same is true, *a fortiori*, in mathematics. The mathematician similarly cannot retain pell-mell all the facts that are presented to him, the more so that it is himself—I was almost going to say his own caprice—that creates these facts. It is he who assembles the elements and constructs a new combination from top to bottom; it is generally not brought to him ready-made by nature.

No doubt it is sometimes the case that a mathematician attacks a problem to satisfy some requirement of physics, that the physicist or the engineer asks him to make a calculation in view of some particular application. Will it be said that we geometricians are to confine ourselves to waiting for orders, and, instead of cultivating our science for our own pleasure, to have no other care but that of accommodating ourselves to our clients' tastes? If the only object of mathematics is to come to the help of those who make a study of nature, it is to them we must look for the word of command. Is this the correct view of the matter? Certainly not; for if we had not cultivated the exact sciences for themselves, we should never have created the mathematical instrument, and when the word of command came from the physicist we should have been found without arms.

Similarly, physicists do not wait to study a phenomenon until some pressing need of material life makes it an absolute necessity, and they are quite right. If the scientists of the eighteenth century had disregarded electricity, because it appeared to them merely a curiosity having no practical interest, we should not have, in the twentieth century, either telegraphy or electro-chemistry or electro-traction. Physicists forced to select are not guided in their selection solely by utility. What method, then, do they pursue in making a selection between the different natural facts? I have explained this in the preceding chapter. The facts that interest them are those that may lead to the discovery of a law, those that have an analogy with many other facts and do not appear to us as isolated, but as closely grouped with others. The isolated fact attracts the attention of all, of the layman as well as the scientist. But what the true scientist alone can see is the link that unites several facts which have a deep but hidden analogy. The anecdote of Newton's apple is probably not true, but it is symbolical, so we will treat it as if it were true. Well, we must suppose that before Newton's day many men had seen apples fall, but none had been able to draw any conclusion. Facts would be barren if there were not minds capable of selecting between them and distinguishing those which have something hidden behind them and recognizing what is hidden—minds which, behind the bare fact, can detect the soul of the fact.

In mathematics we do exactly the same thing. Of the various elements at our disposal we can form millions of different combinations, but any one of these combinations, so long as it is isolated, is absolutely without value; often we have taken great trouble to construct it, but it is of absolutely no use, unless it be, perhaps, to supply a subject for an exercise in secondary schools. It will be quite different as soon as this combination takes its place in a class of analogous combinations whose analogy we have recognized; we shall then be no longer in presence of a fact, but of a law. And then the true discoverer will not be the workman who has patiently built up some of these combinations, but the man who has brought out their relation. The former has only seen the bare fact, the latter alone has detected the soul of the fact. The invention of a new word will often be sufficient to bring out the relation, and the word will be creative. The history of science furnishes us with a host of examples that are familiar to all.

The celebrated Viennese philosopher Mach has said that the part of science is to effect economy of thought, just as a machine effects economy of effort, and this is very true. The savage calculates on his fingers, or by putting together pebbles. By teaching children the multiplication table we save them later on countless operations with pebbles. Some one once recognized, whether by pebbles or otherwise, that 6 times 7 are 42, and had the idea of recording the result, and that is the reason why we do not need to repeat the operation. His time was not wasted even if he was only calculating for his own amusement. His operation only took him two minutes, but it would have taken two million, if a million people had had to repeat it after him.

Thus the importance of a fact is measured by the return it gives—that is, by the amount of thought it enables us to economize.

In physics, the facts which give a large return are those which take their place in a very general law, because they enable us to foresee a very large number of others, and it is exactly the same in mathematics. Suppose I apply myself to a complicated calculation and with much difficulty arrive at a result, I shall have gained nothing by my trouble if it has not enabled me to foresee the results of other analogous calculations, and to direct them with certainty, avoiding the blind groping with which I had to be contented the first time. On the contrary, my time will not have been lost if this very groping has succeeded in revealing to me the profound analogy between the problem just dealt with and a much more extensive class of other problems; if it has shown me at once their resemblances and their differences; if, in a word, it has enabled me to perceive the possibility of a generalization. Then it will not be merely a new result that I have acquired, but a new force.

An algebraical formula which gives us the solution of a type of numerical problems, if we finally replace the letters by numbers, is the simple example which occurs to one's mind at once. Thanks to the formula, a single algebraical calculation saves us the trouble of a constant repetition of numerical calculations. But this is only a rough example: every one feels that there are analogies which cannot be expressed by a formula, and that they are the most valuable.

If a new result is to have any value, it must unite elements long since known, but till then scattered and seemingly foreign to each other, and suddenly introduce order where the appearance of disorder reigned. Then it enables us to see at a glance each of these elements in the place it occupies in the whole. Not only is the new fact valuable on its own account, but it alone gives a value to the old facts it unites. Our mind is frail as our senses are; it would lose itself in the complexity of the world if that complexity were not harmonious; like the short-sighted, it would only see the details, and would be obliged to forget each of these details before examining the next, because it would be incapable of taking in the whole. The only facts worthy of our attention are those which introduce order into this complexity and so make it accessible to us.

Mathematicians attach a great importance to the elegance of their methods and of their results, and this is not mere dilettantism. What is it that gives us the feeling of elegance in a solution or a demonstration? It is the harmony of the different parts, their symmetry, and their happy adjustment; it is, in a word, all that introduces order, all that gives them unity, that enables us to obtain a clear comprehension of the whole as well as of the parts. But that is also precisely what causes it to give a large return; and in fact the more we see this whole clearly and at a single glance, the better we shall perceive the analogies with other neighbouring objects, and consequently the better chance we shall have of guessing the possible generalizations. Elegance may result from the feeling of surprise caused by the unlooked-for occurrence together of objects not habitually associated. In this, again, it is fruitful, since it thus discloses relations till then unrecognized. It is also fruitful even when it only results from the contrast between the simplicity of the means and the complexity of the problem presented, for it then causes us to reflect on the reason for this contrast, and generally shows us that this reason is not chance, but is to be found in some unsuspected law. Briefly stated, the sentiment of mathematical elegance is nothing but the satisfaction due to some conformity between the solution we wish to discover and the necessities of our mind, and it is on account of this very conformity that the solution can be an instrument for us. This æsthetic satisfaction is consequently connected with the economy of thought. Again the comparison with the Erechtheum occurs to me, but I do not wish to serve it up too often.

It is for the same reason that, when a somewhat lengthy calculation has conducted us to some simple and striking result, we are not satisfied until we have shown that we might have foreseen, if not the whole result, at least its most characteristic features. Why is this? What is it that prevents our being contented with a calculation which has taught us apparently all that we wished to know? The reason is that, in analogous cases, the lengthy calculation might not be able to be used again, while this is not true of the reasoning, often semi-intuitive, which might have enabled us to foresee the result. This reasoning being short, we can see all the parts at a single glance, so that we perceive immediately what must be changed to adapt it to all the problems of a similar nature that may be presented. And since it enables us to foresee whether the solution of these problems will be simple, it shows us at least whether the calculation is worth undertaking.

What I have just said is sufficient to show how vain it would be to attempt to replace the mathematician's free initiative by a mechanical process of any kind. In order to obtain a result having any real value, it is not enough to grind out calculations, or to have a machine for putting things in order: it is not order only, but unexpected order, that has a value. A machine can take hold of the bare fact, but the soul of the fact will always escape it.

Since the middle of last century, mathematicians have become more and more anxious to attain to absolute exactness. They are quite right, and this tendency will become more and more marked. In mathematics, exactness is not everything, but without it there is nothing: a demonstration which lacks exactness is nothing at all. This is a truth that I think no one will dispute, but if it is taken too literally it leads us to the conclusion that before 1820, for instance, there was no such thing as mathematics, and this is clearly an exaggeration. The geometricians of that day were willing to assume what we explain by prolix dissertations. This does not mean that they did not see it at all, but they passed it over too hastily, and, in order to see it clearly, they would have had to take the trouble to state it.

Only, is it always necessary to state it so many times? Those who were the first to pay special attention to exactness have given us reasonings that we may attempt to imitate; but if the demonstrations of the future are to be constructed on this model, mathematical works will become exceedingly long, and if I dread length, it is not only because I am afraid of the congestion of our libraries, but because I fear that as they grow in length our demonstrations will lose that appearance of harmony which plays such a useful part, as I have just explained.

It is economy of thought that we should aim at, and therefore it is not sufficient to give models to be copied. We must enable those that come after us to do without the models, and not to repeat a previous reasoning, but summarize it in a few lines. And this has already been done successfully in certain cases. For instance, there was a whole class of reasonings that resembled each other, and were found everywhere; they were perfectly exact, but they were long. One day some one thought of the term "uniformity of convergence," and this term alone made them useless; it was no longer necessary to repeat them, since they could now be assumed. Thus the hair-splitters can render us a double service, first by teaching us to do as they do if necessary, but more especially by enabling us as often as possible not to do as they do, and yet make no sacrifice of exactness.

One example has just shown us the importance of terms in mathematics; but I could quote many others. It is hardly possible to believe what economy of thought, as Mach used to say, can be effected by a well-chosen term. I think I have already said somewhere that mathematics is the art of giving the same name to different things. It is enough that these things, though differing in matter, should be similar in form, to permit of their being, so to speak, run in the same mould. When language has been well chosen, one is astonished to find that all demonstrations made for a known object apply immediately to many new objects: nothing requires to be changed, not even the terms, since the names have become the same.

A well-chosen term is very often sufficient to remove the exceptions permitted by the rules as stated in the old phraseology. This accounts for the invention of negative quantities, imaginary quantities, decimals to infinity, and I know not what else. And we must never forget that exceptions are pernicious, because they conceal laws.

This is one of the characteristics by which we recognize facts which give a great return: they are the facts which permit of these happy innovations of language. The bare fact, then, has sometimes no great interest: it may have been noted many times without rendering any great service to science; it only acquires a value when some more careful thinker perceives the connexion it brings out, and symbolizes it by a term.

The physicists also proceed in exactly the same way. They have invented the term "energy," and the term has been enormously fruitful, because it also creates a law by eliminating exceptions; because it gives the same name to things which differ in matter, but are similar in form.

Among the terms which have exercised the most happy influence I would note "group" and "invariable." They have enabled us to perceive the essence of many mathematical reasonings, and have shown us in how many cases the old mathematicians were dealing with groups without knowing it, and how, believing themselves far removed from each other, they suddenly found themselves close together without understanding why.

To-day we should say that they had been examining isomorphic groups. We now know that, in a group, the matter is of little interest, that the form only is of importance, and that when we are well acquainted with one group, we know by that very fact all the isomorphic groups. Thanks to the terms "group" and "isomorphism," which sum up this subtle rule in a few syllables, and make it readily familiar to all minds, the passage is immediate, and can be made without expending any effort of thinking. The idea of group is, moreover, connected with that of transformation. Why do we attach so much value to the discovery of a new transformation? It is because, from a single theorem, it enables us to draw ten or twenty others. It has the same value as a zero added to the right of a whole number.

This is what has determined the direction of the movement of mathematical science up to the present, and it is also most certainly what will determine it in the future. But the nature of the problems which present themselves contributes to it in an equal degree.

We cannot forget what our aim should be, and in my opinion this aim is a double one. Our science borders on both philosophy and physics, and it is for these two neighbours that we must work. And so we have always seen, and we shall still see, mathematicians advancing in two opposite directions.

On the one side, mathematical science must reflect upon itself, and this is useful because reflecting upon itself is reflecting upon the human mind which has created it; the more so because, of all its creations, mathematics is the one for which it has borrowed least from outside. This is the reason for the utility of certain mathematical speculations, such as those which have in view the study of postulates, of unusual geometries, of functions with strange behaviour. The more these speculations depart from the most ordinary conceptions, and, consequently, from nature and applications to natural problems, the better will they show us what the human mind can do when it is more and more withdrawn from the tyranny of the exterior world; the better, consequently, will they make us know this mind itself.

But it is to the opposite side, to the side of nature, that we must direct our main forces.

There we meet the physicist or the engineer, who says, "Will you integrate this differential equation for me; I shall need it within a week for a piece of construction work that has to be completed by a certain date?" "This equation," we answer, "is not included in one of the types that can be integrated, of which you know there are not very many." "Yes, I know; but, then, what good are you?" More often than not a mutual understanding is sufficient. The engineer does not really require the integral in finite terms, he only requires to know the general behaviour of the integral function, or he merely wants a certain figure which would be easily deduced from this integral if we knew it. Ordinarily we do not know it, but we could calculate the figure without it, if we knew just what figure and what degree of exactness the engineer required.

Formerly an equation was not considered to have been solved until the solution had been expressed by means of a finite number of known functions. But this is impossible in about ninety-nine cases out of a hundred. What we can always do, or rather what we should always try to do, is to solve the problem *qualitatively*, so to speak—that is, to try to know approximately the general form of the curve which represents the unknown function.

It then remains to find the *exact* solution of the problem. But if the unknown cannot be determined by a finite calculation, we can always represent it by an infinite converging series which enables us to calculate it. Can this be regarded as a true solution? The story goes that Newton once communicated to Leibnitz an anagram somewhat like the following: *aaaaabbbeeeeii*, etc. Naturally, Leibnitz did not understand it at all, but we who have the key know that the anagram, translated into modern phraseology, means, "I know how to integrate all differential equations," and we are tempted to make the comment that Newton was either exceedingly fortunate or that he had very singular illusions. What he meant to say was simply that he could form (by means of indeterminate coëfficients) a series of powers formally satisfying the equation presented.

To-day a similar solution would no longer satisfy us, for two reasons—because the convergence is too slow, and because the terms succeed one another without obeying any law. On the other hand the series Θ appears to us to leave nothing to be desired, first, because it converges very rapidly (this is for the practical man who wants his number as quickly as possible), and secondly, because we perceive at a glance the law of the terms, which satisfies the æsthetic requirements of the theorist.

There are, therefore, no longer some problems solved and others unsolved, there are only problems *more or less* solved, according as this is accomplished by a series of more or less rapid convergence or regulated by a more or less harmonious law. Nevertheless an imperfect solution may happen to lead us towards a better one.

Sometimes the series is of such slow convergence that the calculation is impracticable, and we have only succeeded in demonstrating the possibility of the problem. The engineer considers this absurd, and he is right, since it will not help him to complete his construction within the time allowed. He doesn't trouble himself with the question whether it will be of use to the engineers of the twenty-second century. We think differently, and we are sometimes more pleased at having economized a day's work for our grandchildren than an hour for our contemporaries.

Sometimes by groping, so to speak, empirically, we arrive at a formula that is sufficiently convergent.

What more would you have? says the engineer; and yet, in spite of everything, we are not satisfied, for we should have liked to be able to *predict* the convergence. And why? Because if we had known how to predict it in the one case, we should know how to predict it in another. We have been successful, it is true, but that is little in our eyes if we have no real hope of repeating our success.

In proportion as the science develops, it becomes more difficult to take it in in its entirety. Then an attempt is made to cut it in pieces and to be satisfied with one of these pieces—in a word, to specialize. Too great a movement in this direction would constitute a serious obstacle to the progress of the science. As I have said, it is by unexpected concurrences between its different parts that it can make progress. Too much specializing would prohibit these concurrences. Let us hope that congresses, such as those of Heidelberg and Rome, by putting us in touch with each other, will open up a view of our neighbours' territory, and force us to compare it with our own, and so escape in a measure from our own little village. In this way they will be the best remedy against the danger I have just noted.

But I have delayed too long over generalities; it is time to enter into details.

Let us review the different particular sciences which go to make up mathematics; let us see what each of them has done, in what direction it is tending, and what we may expect of it. If the preceding views are correct, we should see that the great progress of the past has been made when two of these sciences have been brought into conjunction, when men have become aware of the similarity of their form in spite of the dissimilarity of their matter, when they have modelled themselves upon each other in such a way that each could profit by the triumphs of the other. At the same time we should look to concurrences of a similar nature for progress in the future.

ARITHMETIC.

The progress of arithmetic has been much slower than that of algebra and analysis, and it is easy to understand the reason. The feeling of continuity is a precious guide which fails the arithmetician. Every whole number is separated from the rest, and has, so to speak, its own individuality; each of them is a sort of exception, and that is the reason why general theorems will always be less common in the theory of numbers, and also why those that do exist will be more hidden and will longer escape detection.

If arithmetic is backward as compared with algebra and analysis, the best thing for it to do is to try to model itself on these sciences, in order to profit by their advance. The arithmetician then should be guided by the analogies with algebra. These analogies are numerous, and if in many cases they have not yet been studied sufficiently closely to become serviceable, they have at least been long foreshadowed, and the very language of the two sciences shows that they have been perceived. Thus we speak of transcendental numbers, and so become aware of the fact that the future classification of these numbers has already a model in the classification of transcendental functions. However, it is not yet very clear how we are to pass from one classification to the other; but if it were clear it would be already done, and would no longer be the work of the future.

The first example that comes to my mind is the theory of congruents, in which we find a perfect parallelism with that of algebraic equations. We shall certainly succeed in completing this parallelism, which must exist, for instance, between the theory of algebraic curves and that of congruents with two variables. When the problems relating to congruents with several variables have been solved, we shall have made the first step towards the solution of many questions of indeterminate analysis.

ALGEBRA.

The theory of algebraic equations will long continue to attract the attention of geometricians, the sides by which it may be approached being so numerous and so different.

It must not be supposed that algebra is finished because it furnishes rules for forming all possible combinations; it still remains to find interesting combinations, those that satisfy such and such conditions. Thus there will be built up a kind of indeterminate analysis, in which the unknown quantities will no longer be whole numbers but polynomials. So this time it is algebra that will model itself on arithmetic, being guided by the analogy of the whole number, either with the whole polynomial with indefinite coefficients, or with the whole polynomial with whole coefficients.

GEOMETRY.

It would seem that geometry can contain nothing that is not already contained in algebra or analysis, and that geometric facts are nothing but the facts of algebra or analysis expressed in another language. It might be supposed, then, that after the review that has just been made, there would be nothing left to say having any special bearing on geometry. But this would imply a failure to recognize the great importance of a well-formed language, or to understand what is added to things themselves by the method of expressing, and consequently of grouping, those things.

To begin with, geometric considerations lead us to set ourselves new problems. These are certainly, if you will, analytical problems, but they are problems we should never have set ourselves on the score of analysis. Analysis, however, profits by them, as it profits by those it is obliged to solve in order to satisfy the requirements of physics.

One great advantage of geometry lies precisely in the fact that the senses can come to the assistance of the intellect, and help to determine the road to be followed, and many minds prefer to reduce the problems of analysis to geometric form. Unfortunately our senses cannot carry us very far, and they leave us in the lurch as soon as we wish to pass outside the three classical dimensions. Does this mean that when we have left this restricted domain in which they would seem to wish to imprison us, we must no longer count on anything but pure analysis, and that all geometry of more than three dimensions is vain and without object? In the generation which preceded ours, the greatest masters would have answered "Yes." To-day we are so familiar with this notion that we can speak of it, even in a university course, without exciting too much astonishment.

But of what use can it be? This is easy to see. In the first place it gives us a very convenient language, which expresses in very concise terms what the ordinary language of analysis would state in long-winded phrases. More than that, this language causes us to give the same name to things which resemble one another, and states analogies which it does not allow us to forget. It thus enables us still to find our way in that space which is too great for us, by calling to our mind continually the visible space, which is only an imperfect image of it, no doubt, but still an image. Here again, as in all the preceding examples, it is the analogy with what is simple that enables us to understand what is complex.

This geometry of more than three dimensions is not a simple analytical geometry, it is not purely quantitative, but also qualitative, and it is principally on this ground that it becomes interesting. There is a science called *Geometry of Position*, which has for its object the study of the relations of position of the different elements of a figure, after eliminating their magnitudes. This geometry is purely qualitative; its theorems would remain true if the figures, instead of being exact, were rudely imitated by a child. We can also construct a *Geometry of Position* ot more than three dimensions. The importance of *Geometry of Position* is immense, and I cannot insist upon it too much; what Riemann, one of its principal creators, has gained from it would be sufficient to demonstrate this. We must succeed in constructing it completely in the higher spaces, and we shall then have an instrument which will enable us really to see into hyperspace and to supplement our senses.

The problems of *Geometry of Position* would perhaps not have presented themselves if only the language of analysis had been used. Or rather I am wrong, for they would certainly have presented themselves, since their solution is necessary for a host of questions of analysis, but they would have presented themselves isolated, one after the other, and without our being able to perceive their common link.

CANTORISM.

I have spoken above of the need we have of returning continually to the first principles of our science, and of the advantage of this process to the study of the human mind. It is this need which has inspired two attempts which have held a very great place in the most recent history of mathematics. The first is Cantorism, and the services it has rendered to the science are well known. Cantor introduced into the science a new method of considering mathematical infinity, and I shall have occasion to speak of it again in Book II., chapter iii. One of the characteristic features of Cantorism is that, instead of rising to the general by erecting more and more complicated constructions, and defining by construction, it starts with the *genus supremum* and only defines, as the scholastics would have said, *per genus proximum et differentiam specificam.* Hence the horror he has sometimes inspired in certain minds, such as Hermitte's, whose favourite idea was to compare the mathematical with the natural sciences. For the greater number of us these prejudices had been dissipated, but it has come about that we have run against certain paradoxes and apparent contradictions, which would have rejoiced the heart of Zeno of Elea and the school of Megara. Then began the business of searching for a remedy, each man his own way. For my part I think, and I am not alone in so thinking, that the important thing is never to introduce any entities but such as can be completely defined in a finite number of words. Whatever be the remedy adopted, we can promise ourselves the joy of the doctor called in to follow a fine pathological case.

The Search for Postulates.

Attempts have been made, from another point of view, to enumerate the axioms and postulates more or less concealed which form the foundation of the different mathematical theories, and in this direction Mr. Hilbert has obtained the most brilliant results. It seems at first that this domain must be strictly limited, and that there will be nothing more to do when the inventory has been completed, which cannot be long. But when everything has been enumerated, there will be many ways of classifying it all. A good librarian always finds work to do, and each new classification will be instructive for the philosopher.

I here close this review, which I cannot dream of making complete. I think that these examples will have been sufficient to show the mechanism by which the mathematical sciences have progressed in the past, and the direction in which they must advance in the future.

III.

MATHEMATICAL DISCOVERY.

THE genesis of mathematical discovery is a problem which must inspire the psychologist with the keenest interest. For this is the process in which the human mind seems to borrow least from the exterior world, in which it acts, or appears to act, only by itself and on itself, so that by studying the process of geometric thought we may hope to arrive at what is most essential in the human mind.

This has long been understood, and a few months ago a review called *l'Enseignement Mathématique*, edited by MM. Laisant and Fehr, instituted an enquiry into the habits of mind and methods of work of different mathematicians. I had outlined the principal features of this article when the results of the enquiry were published, so that I have hardly been able to make any use of them, and I will content myself with saying that the majority of the evidence confirms my conclusions. I do not say there is unanimity, for on an appeal to universal suffrage we cannot hope to obtain unanimity.

One first fact must astonish us, or rather would astonish us if we were not too much accustomed to it. How does it happen that there are people who do not understand mathematics? If the science invokes only the rules of logic, those accepted by all well-formed minds, if its evidence is founded on principles that are common to all men, and that none but a madman would attempt to deny, how does it happen that there are so many people who are entirely impervious to it?

There is nothing mysterious in the fact that every one is not capable of discovery. That every one should not be able to retain a demonstration he has once learnt is still comprehensible. But what does seem most surprising, when we consider it, is that any one should be unable to understand a mathematical argument at the very moment it is stated to him. And yet those who can only follow the argument with difficulty are in a majority; this is incontestable, and the experience of teachers of secondary education will certainly not contradict me.

And still further, how is error possible in mathematics? A healthy intellect should not be guilty of any error in logic, and yet there are very keen minds which will not make a false step in a short argument such as those we have to make in the ordinary actions of life, which yet are incapable of following or repeating without error the demonstrations of mathematics which are longer, but which are, after all, only accumulations of short arguments exactly analogous to those they make so easily. Is it necessary to add that mathematicians themselves are not infallible?

The answer appears to me obvious. Imagine a long series of syllogisms in which the conclusions of those that precede form the premises of those that follow. We shall be capable of grasping each of the syllogisms, and it is not in the passage from premises to conclusion that we are in danger of going astray. But between the moment when we meet a proposition for the first time as the conclusion of one syllogism, and the moment when we find it once more as the premise of another syllogism, much time will sometimes have elapsed, and we shall have unfolded many links of the chain; accordingly it may well happen that we shall have forgotten it, or, what is more serious, forgotten its meaning. So we may chance to replace it by a somewhat different proposition, or to preserve the same statement but give it a slightly different meaning, and thus we are in danger of falling into error.

A mathematician must often use a rule, and, naturally, he begins by demonstrating the rule. At the moment the demonstration is quite fresh in his memory he understands perfectly its meaning and significance, and he is in no danger of changing it. But later on he commits it to memory, and only applies it in a mechanical way, and then, if his memory fails him, he may apply it wrongly. It is thus, to take a simple and almost vulgar example, that we sometimes make mistakes in calculation, because we have forgotten our multiplication table.

On this view special aptitude for mathematics would be due to nothing but a very certain memory or a tremendous power of attention. It would be a quality analogous to that of the whist player who can remember the cards played, or, to rise a step higher, to that of the chess player who can picture a very great number of combinations and retain them in his memory. Every good mathematician should also be a good chess player and *vice versâ*, and similarly he should be a good numerical calculator. Certainly this sometimes happens, and thus Gauss was at once a geometrician of genius and a very precocious and very certain calculator.

But there are exceptions, or rather I am wrong, for I cannot call them exceptions, otherwise the exceptions would be more numerous than the cases of conformity with the rule. On the contrary, it was Gauss who was an exception. As for myself, I must confess I am absolutely incapable of doing an addition sum without a mistake. Similarly I should be a very bad chess player. I could easily calculate that by playing in a certain way I should be exposed to such and such a danger; I should then review many other moves, which I should reject for other reasons, and I should end by making the move I first examined, having forgotten in the interval the danger I had foreseen.

In a word, my memory is not bad, but it would be insufficient to make me a good chess player. Why, then, does it not fail me in a difficult mathematical argument in which the majority of chess players would be lost? Clearly because it is guided by the general trend of the argument. A mathematical demonstration is not a simple juxtaposition of syllogisms; it consists of syllogisms *placed in a certain order*, and the order in which these elements are placed is much more important than the elements themselves. If I have the feeling, so to speak the intuition, of this order, so that I can perceive the whole of the argument at a glance, I need no longer be afraid of forgetting one of the elements; each of them will place itself naturally in the position prepared for it, without my having to make any effort of memory.

It seems to me, then, as I repeat an argument I have learnt, that I could have discovered it. This is often only an illusion; but even then, even if I am not clever enough to create for myself, I rediscover it myself as I repeat it.

We can understand that this feeling, this intuition of mathematical order, which enables us to guess hidden harmonies and relations, cannot belong to every one. Some have neither this delicate feeling that is difficult to define, nor a power of memory and attention above the common, and so they are absolutely incapable of understanding even the first steps of higher mathematics. This applies to the majority of people. Others have the feeling only in a slight degree, but they are gifted with an uncommon memory and a great capacity for attention. They learn the details one after the other by heart, they can understand mathemathics and sometimes apply them, but they are not in a condition to create. Lastly, others possess the special intuition I have spoken of more or less highly developed, and they can not only understand mathematics, even though their memory is in no way extraordinary, but they can become creators, and seek to make discovery with more or less chance of success, according as their intuition is more or less developed.

What, in fact, is mathematical discovery? It does not consist in making new combinations with mathematical entities that are already known. That can be done by any one, and the combinations that could be so formed would be infinite in number, and the greater part of them would be absolutely devoid of interest. Discovery consists precisely in not constructing useless combinations, but in constructing those that are useful, which are an infinitely small minority. Discovery is discernment, selection.

How this selection is to be made I have explained above. Mathematical facts worthy of being studied are those which, by their analogy with other facts, are capable of conducting us to the knowledge of a mathematical law, in the same way that experimental facts conduct us to the knowledge of a physical law. They are those which reveal unsuspected relations between other facts, long since known, but wrongly believed to be unrelated to each other.

Among the combinations we choose, the most fruitful are often those which are formed of elements borrowed from widely separated domains. I do not mean to say that for discovery it is sufficient to bring together objects that are as incongruous as possible. The greater part of the combinations so formed would be entirely fruitless, but some among them, though very rare, are the most fruitful of all.

Discovery, as I have said, is selection. But this is perhaps not quite the right word. It suggests a purchaser who has been shown a large number of samples, and examines them one after the other in order to make his selection. In our case the samples would be so numerous that a whole life would not give sufficient time to examine them. Things do not happen in this way. Unfruitful combinations do not so much as present themselves to the mind of the discoverer. In the field of his consciousness there never appear any but really useful combinations, and some that he rejects, which, however, partake to some extent of the character of useful combinations. Everything happens as if the discoverer were a secondary examiner who had only to interrogate candidates declared eligible after passing a preliminary test.

But what I have said up to now is only what can be observed or inferred by reading the works of geometricians, provided they are read with some reflection.

It is time to penetrate further, and to see what happens in the very soul of the mathematician. For this purpose I think I cannot do better than recount my personal recollections. Only I am going to confine myself to relating how I wrote my first treatise on Fuchsian functions. I must apologize, for I am going to introduce some technical expressions, but they need not alarm the reader, for he has no need to understand them. I shall say, for instance, that I found the demonstration of such and such a theorem under such and such circumstances; the theorem will have a barbarous name that many will not know, but that is of no importance. What is interesting for the psychologist is not the theorem but the circumstances.

For a fortnight I had been attempting to prove that there could not be any function analogous to what I have since called Fuchsian functions. I was at that time very ignorant. Every day I sat down at my table and spent an hour or two trying a great number of combinations, and I arrived at no result. One night I took some black coffee, contrary to my custom, and was unable to sleep. A host of ideas kept surging in my head; I could almost feel then jostling one another, until two of them coalesced, so to speak, to form a stable combination. When morning came, I had established the existence of one class of Fuchsian functions, those that are derived from the hypergeometric series. I had only to verify the results, which only took a few hours.

Then I wished to represent these functions by the quotient of two series. This idea was perfectly conscious and deliberate; I was guided by the analogy with elliptical functions. I asked myself what must be the properties of these series, if they existed, and I succeeded without difficulty in forming the series that I have called Theta-Fuchsian.

At this moment I left Caen, where I was then living, to take part in a geological conference arranged by the School of Mines. The incidents of the journey made me forget my mathematical work. When we arrived at Coutances, we got into a break to go for a drive, and, just as I put my foot on the step, the idea came to me, though nothing in my former thoughts seemed to have prepared me for it, that the transformations I had used to define Fuchsian functions were identical with those of non-Euclidian geometry. I made no verification, and had no time to do so, since I took up the conversation again as soon as I had sat down in the break, but I felt absolute certainty at once. When I got back to Caen I verified the result at my leisure to satisfy my conscience.

I then began to study arithmetical questions without any great apparent result, and without suspecting that they could have the least connexion with my previous researches. Disgusted at my want of success, I went away to spend a few days at the seaside, and thought of entirely different things. One day, as I was walking on the cliff, the idea came to me, again with the same characteristics of conciseness, suddenness, and immediate certainty, that arithmetical transformations of indefinite ternary quadratic forms are identical with those of non-Euclidian geometry.

Returning to Caen, I reflected on this result and deduced its consequences. The example of quadratic forms showed me that there are Fuchsian groups other than those which correspond with the hypergeometric series; I saw that I could apply to them the theory of the Theta-Fuchsian series, and that, consequently, there are Fuchsian functions other than those which are derived from the hypergeometric series, the only ones I knew up to that time. Naturally, I proposed to form all these functions. I laid siege to them systematically and captured all the outworks one after the other. There was one, however, which still held out, whose fall would carry with it that of the central fortress. But all my efforts were of no avail at first, except to make me better understand the difficulty, which was already something. All this work was perfectly conscious.

Thereupon I left for Mont-Valérien, where I had to serve my time in the army, and so my mind was preoccupied with very different matters. One day, as I was crossing the street, the solution of the difficulty which had brought me to a standstill came to me all at once. I did not try to fathom it immediately, and it was only after my service was finished that I returned to the question. I had all the elements, and had only to assemble and arrange them. Accordingly I composed my definitive treatise at a sitting and without any difficulty.

It is useless to multiply examples, and I will content myself with this one alone. As regards my other researches, the accounts I should give would be exactly similar, and the observations related by other mathematicians in the enquiry of *l'Enseignement Mathématique* would only confirm them.

One is at once struck by these appearances of sudden illumination, obvious indications of a long course of previous unconscious work. The part played by this unconscious work in mathematical discovery seems to me indisputable, and we shall find traces of it in other cases where it is less evident. Often when a man is working at a difficult question, he accomplishes nothing the first time he sets to work. Then he takes more or less of a rest, and sits down again at his table. During the first half-hour he still finds nothing, and then all at once the decisive idea presents itself to his mind. We might say that the conscious work proved more fruitful because it was interrupted and the rest restored force and freshness to the mind. But it is more probable that the rest was occupied with unconscious work, and that the result of this work was afterwards revealed to the geometrician exactly as in the cases I have quoted, except that the revelation, instead of coming to light during a walk or a journey, came during a period of conscious work, but independently of that work, which at most only performs the unlocking process, as if it were the spur that excited into conscious form the results already acquired during the rest, which till then remained unconscious.

There is another remark to be made regarding the conditions of this unconscious work, which is, that it is not possible, or in any case not fruitful, unless it is first preceded and then followed by a period of conscious work. These sudden inspirations are never produced (and this is sufficiently proved already by the examples I have quoted) except after some days of voluntary efforts which appeared absolutely fruitless, in which one thought one had accomplished nothing, and seemed to be on a totally wrong track. These efforts, however, were not as barren as one thought; they set the unconscious machine in motion, and without them it would not have worked at all, and would not have produced anything.

The necessity for the second period of conscious work can be even more readily understood. It is necessary to work out the results of the inspiration, to deduce the immediate consequences and put them in order and to set out the demonstrations; but, above all, it is necessary to verify them. I have spoken of the feeling of absolute certainty which accompanies the inspiration; in the cases quoted this feeling was not deceptive, and more often than not this will be the case. But we must beware of thinking that this is a rule without exceptions. Often the feeling deceives us without being any less distinct on that account, and we only detect it when we attempt to establish the demonstration. I have observed this fact most notably with regard to ideas that have come to me in the morning or at night when I have been in bed in a semi-somnolent condition.

Such are the facts of the case, and they suggest the following reflections. The result of all that precedes is to show that the unconscious ego, or, as it is called, the subliminal ego, plays a most important part in mathematical discovery. But the subliminal ego is generally thought of as purely automatic. Now we have seen that mathematical work is not a simple mechanical work, and that it could not be entrusted to any machine, whatever the degree of perfection we suppose it to have been brought to. It is not merely a question of applying certain rules, of manufacturing as many combinations as possible according to certain fixed laws. The combinations so obtained would be extremely numerous, useless, and encumbering. The real work of the discoverer consists in choosing between these combinations with a view to eliminating those that are useless, or rather not giving himself the trouble of making them at all. The rules which must guide this choice are extremely subtle and delicate, and it is practically impossible to state them in precise language; they must be felt rather than formulated. Under these conditions, how can we imagine a sieve capable of applying them mechanically?

The following, then, presents itself as a first hypothesis. The subliminal ego is in no way inferior to the conscious ego; it is not purely automatic; it is capable of discernment; it has tact and lightness of touch; it can select, and it can divine. More than that, it can divine better than the conscious ego, since it succeeds where the latter fails. In a word, is not the subliminal ego superior to the conscious ego? The importance of this question will be readily understood. In a recent lecture, M. Boutroux showed how it had arisen on entirely different occasions, and what consequences would be involved by an answer in the affirmative. (See also the same author's *Science et Religion*, pp. 313 *et seq.*)

Are we forced to give this affirmative answer by the facts I have just stated? I confess that, for my part, I should be loth to accept it. Let us, then, return to the facts, and see if they do not admit of some other explanation.

It is certain that the combinations which present themselves to the mind in a kind of sudden illumination after a somewhat prolonged period of unconscious work are generally useful and fruitful combinations, which appear to be the result of a preliminary sifting. Does it follow from this that the subliminal ego, having divined by a delicate intuition that these combinations could be useful, has formed none but these, or has it formed a great many others which were devoid of interest, and remained unconscious?

Under this second aspect, all the combinations are formed as a result of the automatic action of the subliminal ego, but those only which are interesting find their way into the field of consciousness. This, too, is most mysterious. How can we explain the fact that, of the thousand products of our unconscious activity, some are invited to cross the threshold, while others remain outside? Is it mere chance that gives them this privilege? Evidently not. For instance, of all the excitements of our senses, it is only the most intense that retain our attention, unless it has been directed upon them by other causes. More commonly the privileged unconscious phenomena, those that are capable of becoming conscious, are those which, directly or indirectly, most deeply affect our sensibility.

It may appear surprising that sensibility should be introduced in connexion with mathematical demonstrations, which, it would seem, can only interest the intellect. But not if we bear in mind the feeling of mathematical beauty, of the harmony of numbers and forms and of geometric elegance. It is a real æsthetic feeling that all true mathematicians recognize, and this is truly sensibility.

Now, what are the mathematical entities to which we attribute this character of beauty and elegance, which are capable of developing in us a kind of æsthetic emotion? Those whose elements are harmoniously arranged so that the mind can, without effort, take in the whole without neglecting the details. This harmony is at once a satisfaction to our æsthetic requirements, and an assistance to the mind which it supports and guides. At the same time, by setting before our eyes a well-ordered whole, it gives us a presentiment of a mathematical law. Now, as I have said above, the only mathematical facts worthy of retaining our attention and capable of being useful are those which can make us acquainted with a mathematical law. Accordingly we arrive at the following conclusion. The useful combinations are precisely the most beautiful, I mean those that can most charm that special sensibility that all mathematicians know, but of which laymen are so ignorant that they are often tempted to smile at it.

What follows, then? Of the very large number of combinations which the subliminal ego blindly forms, almost all are without interest and without utility. But, for that very reason, they are without action on the æsthetic sensibility; the consciousness will never know them. A few only are harmonious, and consequently at once useful and beautiful, and they will be capable of affecting the geometrician's special sensibility I have been speaking of; which, once aroused, will direct our attention upon them, and will thus give them the opportunity of becoming conscious.

This is only a hypothesis, and yet there is an observation which tends to confirm it. When a sudden illumination invades the mathematician's mind, it most frequently happens that it does not mislead him. But it also happens sometimes, as I have said, that it will not stand the test of verification. Well, it is to be observed almost always that this false idea, if it had been correct, would have flattered our natural instinct for mathematical elegance.

Thus it is this special æsthetic sensibility that plays the part of the delicate sieve of which I spoke above, and this makes it sufficiently clear why the man who has it not will never be a real discoverer.

All the difficulties, however, have not disappeared. The conscious ego is strictly limited, but as regards the subliminal ego, we do not know its limitations, and that is why we are not too loth to suppose that in a brief space of time it can form more different combinations than could be comprised in the whole life of a conscient being. These limitations do exist, however. Is it conceivable that it can form all the possible combinations, whose number staggers the imagination? Nevertheless this would seem to be necessary, for if it produces only a small portion of the combinations, and that by chance, there will be very small likelihood of the *right* one, the one that must be selected, being found among them.

Perhaps we must look for the explanation in that period of preliminary conscious work which always precedes all fruitful unconscious work. If I may be permitted a crude comparison, let us represent the future elements of our combinations as something resembling Epicurus's hooked atoms. When the mind is in complete repose these atoms are immovable; they are, so to speak, attached to the wall. This complete repose may continue indefinitely without the atoms meeting, and, consequently, without the possibility of the formation of any combination.

On the other hand, during a period of apparent repose, but of unconscious work, some of them are detached from the wall and set in motion. They plough through space in all directions, like a swarm of gnats, for instance, or, if we prefer a more learned comparison, like the gaseous molecules in the kinetic theory of gases. Their mutual collisions may then produce new combinations.

What is the part to be played by the preliminary conscious work? Clearly it is to liberate some of these atoms, to detach them from the wall and set them in motion. We think we have accomplished nothing, when we have stirred up the elements in a thousand different ways to try to arrange them, and have not succeeded in finding a satisfactory arrangement. But after this agitation imparted to them by our will, they do not return to their original repose, but continue to circulate freely.

Now our will did not select them at random, but in pursuit of a perfectly definite aim. Those it has liberated are not, therefore, chance atoms; they are those from which we may reasonably expect the desired solution. The liberated atoms will then experience collisions, either with each other, or with the atoms that have remained stationary, which they will run against in their course. I apologize once more. My comparison is very crude, but I cannot well see how I could explain my thought in any other way.

However it be, the only combinations that have any chance of being formed are those in which one at least of the elements is one of the atoms deliberately selected by our will. Now it is evidently among these that what I called just now the *right* combination is to be found. Perhaps there is here a means of modifying what was paradoxical in the original hypothesis.

Yet another observation. It never happens that unconscious work supplies *ready-made* the result of a lengthy calculation in which we have only to apply fixed rules. It might be supposed that the subliminal ego, purely automatic as it is, was peculiarly fitted for this kind of work, which is, in a sense, exclusively mechanical. It would seem that, by thinking overnight of the factors of a multiplication sum, we might hope to find the product ready-made for us on waking; or, again, that an algebraical calculation, for instance, or a verification could be made unconsciously. Observation proves that such is by no means the case. All that we can hope from these inspirations, which are the fruits of unconscious work, is to obtain points of departure for such calculations. As for the calculations themselves, they must be made in the second period of conscious work which follows the inspiration, and in which the results of the inspiration are verified and the consequences deduced. The rules of these calculations are strict and complicated; they demand discipline, attention, will, and consequently consciousness. In the subliminal ego, on the contrary, there reigns what I would call liberty, if one could give this name to the mere absence of discipline and to disorder born of chance. Only, this very disorder permits of unexpected couplings.

I will make one last remark. When I related above some personal observations, I spoke of a night of excitement, on which I worked as though in spite of myself. The cases of this are frequent, and it is not necessary that the abnormal cerebral activity should be caused by a physical stimulant, as in the case quoted. Well, it appears that, in these cases, we are ourselves assisting at our own unconscious work, which becomes partly perceptible to the over-excited consciousness, but does not on that account change its nature. We then become vaguely aware of what distinguishes the two mechanisms, or, if you will, of the methods of working of the two egos. The psychological observations I have thus succeeded in making appear to me, in their general characteristics, to confirm the views I have been enunciating.

Truly there is great need of this, for in spite of everything they are and remain largely hypothetical. The interest of the question is so great that I do not regret having submitted them to the reader.

IV.

CHANCE.

I.

"HOW can we venture to speak of the laws of chance? Is not chance the antithesis of all law?" It is thus that Bertrand expresses himself at the beginning of his "Calculus of Probabilities." Probability is the opposite of certainty; it is thus what we are ignorant of, and consequently it would seem to be what we cannot calculate. There is here at least an apparent contradiction, and one on which much has already been written.

To begin with, what is chance? The ancients distinguished between the phenomena which seemed to obey harmonious laws, established once for all, and those that they attributed to chance, which were those that could not be predicted because they were not subject to any law. In each domain the precise laws did not decide everything, they only marked the limits within which chance was allowed to move. In this conception, the word chance had a precise, objective meaning; what was chance for one was also chance for the other and even for the gods.

But this conception is not ours. We have become complete determinists, and even those who wish to reserve the right of human free will at least allow determinism to reign undisputed in the inorganic world. Every phenomenon, however trifling it be, has a cause, and a mind infinitely powerful and infinitely well-informed concerning the laws of nature could have foreseen it from the beginning of the ages. If a being with such a mind existed, we could play no game of chance with him; we should always lose.

For him, in fact, the word chance would have no meaning, or rather there would be no such thing as chance. That there is for us is only on account of our frailty and our ignorance. And even without going beyond our frail humanity, what is chance for the ignorant is no longer chance for the learned. Chance is only the measure of our ignorance. Fortuitous phenomena are, by definition, those whose laws we are ignorant of.

But is this definition very satisfactory? When the first Chaldean shepherds followed with their eyes the movements of the stars, they did not yet know the laws of astronomy, but would they have dreamed of saying that the stars move by chance? If a modern physicist is studying a new phenomenon, and if he discovers its law on Tuesday, would he have said on Monday that the phenomenon was fortuitous? But more than this, do we not often invoke what Bertrand calls the laws of chance in order to predict a phenomenon? For instance, in the kinetic theory of gases, we find the well-known laws of Mariotte and of Gay-Lussac, thanks to the hypothesis that the velocities of the gaseous molecules vary irregularly, that is to say, by chance.

The observable laws would be much less simple, say all the physicists, if the velocities were regulated by some simple elementary law, if the molecules were, as they say, *organized*, if they were subject to some discipline. It is thanks to chance—that is to say, thanks to our ignorance, that we can arrive at conclusions. Then if the word chance is merely synonymous with ignorance, what does this mean? Must we translate as follows?—

"You ask me to predict the phenomena that will be produced. If I had the misfortune to know the laws of these phenomena, I could not succeed except by inextricable calculations, and I should have to give up the attempt to answer you; but since I am fortunate enough to be ignorant of them, I will give you an answer at once. And, what is more extraordinary still, my answer will be right."

Chance, then, must be something more than the name we give to our ignorance. Among the phenomena whose causes we are ignorant of, we must distinguish between fortuitous phenomena, about which the calculation of probabilities will give us provisional information, and those that are not fortuitous, about which we can say nothing, so long as we have not determined the laws that govern them. And as regards the fortuitous phenomena themselves, it is clear that the information that the calculation of probabilities supplies will not cease to be true when the phenomena are better known.

The manager of a life insurance company does not know when each of the assured will die, but he relies upon the calculation of probabilities and on the law of large numbers, and he does not make a mistake, since he is able to pay dividends to his shareholders. These dividends would not vanish if a very far-sighted and very indiscreet doctor came, when once the policies were signed, and gave the manager information on the chances of life of the assured. The doctor would dissipate the ignorance of the manager, but he would have no effect upon the dividends, which are evidently not a result of that ignorance.

II.

In order to find the best definition of chance, we must examine some of the facts which it is agreed to regard as fortuitous, to which the calculation of probabilities seems to apply. We will then try to find their common characteristics.

We will select unstable equilibrium as our first example. If a cone is balanced on its point, we know very well that it will fall, but we do not know to which side; it seems that chance alone will decide. If the cone were perfectly symmetrical, if its axis were perfectly vertical, if it were subject to no other force but gravity, it would not fall at all. But the slightest defect of symmetry will make it lean slightly to one side or other, and as soon as it leans, be it ever so little, it will fall altogether to that side. Even if the symmetry is perfect, a very slight trepidation, or a breath of air, may make it incline a few seconds of arc, and that will be enough to determine its fall and even the direction of its fall, which will be that of the original inclination.

A very small cause which escapes our notice determines a considerable effect that we cannot fail to see, and then we say that that effect is due to chance. If we knew exactly the laws of nature and the situation of the universe at the initial moment, we could predict exactly the situation of that same universe at a succeeding moment. But, even if it were the case that the natural laws had no longer any secret for us, we could still only know the initial situation *approximately*. If that enabled us to predict the succeeding situation *with the same approximation*, that is all we require, and we should say that the phenomenon had been predicted, that it is governed by laws. But it is not always so; it may happen that small differences in the initial conditions produce very great ones in the final phenomena. A small error in the former will produce an enormous error in the latter. Prediction becomes impossible, and we have the fortuitous phenomenon.

Our second example will be very much like our first, and we will borrow it from meteorology. Why have meteorologists such difficulty in predicting the weather with any certainty? Why is it that showers and even storms seem to come by chance, so that many people think it quite natural to pray for rain or fine weather, though they would consider it ridiculous to ask for an eclipse by prayer? We see that great disturbances are generally produced in regions where the atmosphere is in unstable equilibrium. The meteorologists see very well that the equilibrium is unstable, that a cyclone will be formed somewhere, but exactly where they are not in a position to say; a tenth of a degree more or less at any given point, and the cyclone will burst here and not there, and extend its ravages over districts it would otherwise have spared. If they had been aware of this tenth of a degree, they could have known it beforehand, but the observations were neither sufficiently comprehensive nor sufficiently precise, and that is the reason why it all seems due to the intervention of chance. Here, again, we find the same contrast between a very trifling cause that is inappreciable to the observer, and considerable effects, that are sometimes terrible disasters.

Let us pass to another example, the distribution of the minor planets on the Zodiac. Their initial longitudes may have had some definite order, but their mean motions were different and they have been revolving for so long that we may say that practically they are distributed *by chance* throughout the Zodiac. Very small initial differences in their distances from the sun, or, what amounts to the same thing, in their mean motions, have resulted in enormous differences in their actual longitudes. A difference of a thousandth part of a second in the mean daily motion will have the effect of a second in three years, a degree in ten thousand years, a whole circumference in three or four millions of years, and what is that beside the time that has elapsed since the minor planets became detached from Laplace's nebula? Here, again, we have a small cause and a great effect, or better, small differences in the cause and great differences in the effect.

The game of roulette does not take us so far as it might appear from the preceding example. Imagine a needle that can be turned about a pivot on a dial divided into a hundred alternate red and black sections. If the needle stops at a red section we win; if not, we lose. Clearly, all depends on the initial impulse we give to the needle. I assume that the needle will make ten or twenty revolutions, but it will stop earlier or later according to the strength of the spin I have given it. Only a variation of a thousandth or a two-thousandth in the impulse is sufficient to determine whether my needle will stop at a black section or at the following section, which is red. These are differences that the muscular sense cannot appreciate, which would escape even more delicate instruments. It is, accordingly, impossible for me to predict what the needle I have just spun will do, and that is why my heart beats and I hope for everything from chance. The difference in the cause is imperceptible, and the difference in the effect is for me of the highest importance, since it affects my whole stake.

III.

In this connexion I wish to make a reflection that is somewhat foreign to my subject. Some years ago a certain philosopher said that the future was determined by the past, but not the past by the future; or, in other words, that from the knowledge of the present we could deduce that of the future but not that of the past; because, he said, one cause can produce only one effect, while the same effect can be produced by several different causes. It is obvious that no scientist can accept this conclusion. The laws of nature link the antecedent to the consequent in such a way that the antecedent is determined by the consequent just as much as the consequent is by the antecedent. But what can have been the origin of the philosopher's error? We know that, in virtue of Carnot's principle, physical phenomena are irreversible and that the world is tending towards uniformity. When two bodies of different temperatures are in conjunction, the warmer gives up heat to the colder, and accordingly we can predict that the temperatures will become equal. But once the temperatures have become equal, if we are asked about the previous state, what can we answer? We can certainly say that one of the bodies was hot and the other cold, but we cannot guess which of the two was formerly the warmer.

And yet in reality the temperatures never arrive at perfect equality. The difference between the temperatures only tends towards zero asymptotically. Accordingly there comes a moment when our thermometers are powerless to disclose it. But if we had thermometers a thousand or a hundred thousand times more sensitive, we should recognize that there is still a small difference, and that one of the bodies has remained a little warmer than the other, and then we should be able to state that this is the one which was formerly very much hotter than the other.

So we have, then, the reverse of what we found in the preceding examples, great differences in the cause and small differences in the effect. Flammarion once imagined an observer moving away from the earth at a velocity greater than that of light. For him time would have its sign changed, history would be reversed, and Waterloo would come before Austerlitz. Well, for this observer effects and causes would be inverted, unstable equilibrium would no longer be the exception; on account of the universal irreversibility, everything would seem to him to come out of a kind of chaos in unstable equilibrium, and the whole of nature would appear to him to be given up to chance.

IV.

We come now to other arguments, in which we shall see somewhat different characteristics appearing, and first let us take the kinetic theory of gases. How are we to picture a receptacle full of gas? Innumerable molecules, animated with great velocities, course through the receptacle in all directions; every moment they collide with the sides or else with one another, and these collisions take place under the most varied conditions. What strikes us most in this case is not the smallness of the causes, but their complexity. And yet the former element is still found here, and plays an important part. If a molecule deviated from its trajectory to left or right in a very small degree as compared with the radius of action of the gaseous molecules, it would avoid a collision, or would suffer it under different conditions, and that would alter the direction of its velocity after the collision perhaps by 90 or 180 degrees.

That is not all. It is enough, as we have just seen, that the molecule should deviate before the collision in an infinitely small degree, to make it deviate after the collision in a finite degree. Then, if the molecule suffers two successive collisions, it is enough that it should deviate before the first collision in a degree of infinite smallness of the second order, to make it deviate after the first collision in a degree of infinite smallness of the first order, and after the second collision in a finite degree. And the molecule will not suffer two collisions only, but a great number each second.

So that if the first collision multiplied the deviation by a very large number, A, after n collisions it will be multiplied by A^n. It will, therefore, have become very great, not only because A is large—that is to say, because small causes produce great effects—but because the exponent n is large, that is to say, because the collisions are very numerous and the causes very complex.

Let us pass to a second example. Why is it that in a shower the drops of rain appear to us to be distributed by chance? It is again because of the complexity of the causes which determine their formation. Ions have been distributed through the atmosphere; for a long time they have been subjected to constantly changing air currents; they have been involved in whirlwinds of very small dimensions, so that their final distribution has no longer any relation to their original distribution. Suddenly the temperature falls, the vapour condenses, and each of these ions becomes the centre of a raindrop. In order to know how these drops will be distributed and how many will fall on each stone of the pavement, it is not enough to know the original position of the ions, but we must calculate the effect of a thousand minute and capricious air currents.

It is the same thing again if we take grains of dust in suspension in water. The vessel is permeated by currents whose law we know nothing of except that it is very complicated. After a certain length of time the grains will be distributed by chance, that is to say uniformly, throughout the vessel, and this is entirely due to the complication of the currents If they obeyed some simple law—if, for instance the vessel were revolving and the currents revolved in circles about its axis—the case would be altered, for each grain would retain its original height and its original distance from the axis.

We should arrive at the same result by picturing the mixing of two liquids or of two fine powders. To take a rougher example, it is also what happens when a pack of cards is shuffled. At each shuffle the cards undergo a permutation similar to that studied in the theory of substitutions. What will be the resulting permutation? The probability that it will be any particular permutation (for instance, that which brings the card occupying the position $\phi(n)$ before the permutation into the position n), this probability, I say, depends on the habits of the player. But if the player shuffles the cards long enough, there will be a great number of successive permutations, and the final order which results will no longer be governed by anything but chance; I mean that all the possible orders will be equally probable. This result is due to the great number of successive permutations, that is to say, to the complexity of the phenomenon.

A final word on the theory of errors. It is a case in which the causes have complexity and multiplicity. How numerous are the traps to which the observer is exposed, even with the best instrument. He must take pains to look out for and avoid the most flagrant, those which give birth to systematic errors. But when he has eliminated these, admitting that he succeeds in so doing, there still remain many which, though small, may become dangerous by the accumulation of their effects. It is from these that accidental errors arise, and we attribute them to chance, because their causes are too complicated and too numerous. Here again we have only small causes, but each of them would only produce a small effect; it is by their union and their number that their effects become formidable.

V.

There is yet a third point of view, which is less important than the two former, on which I will not lay so much stress. When we are attempting to predict a fact and making an examination of the antecedents, we endeavour to enquire into the anterior situation. But we cannot do this for every part of the universe, and we are content with knowing what is going on in the neighbourhood of the place where the fact will occur, or what appears to have some connexion with the fact. Our enquiry cannot be complete, and we must know how to select. But we may happen to overlook circumstances which, at first sight, seemed completely foreign to the anticipated fact, to which we should never have dreamed of attributing any influence, which nevertheless, contrary to all anticipation, come to play an important part.

A man passes in the street on the way to his business. Some one familiar with his business could say what reason he had for starting at such an hour and why he went by such a street. On the roof a slater is at work. The contractor who employs him could, to a certain extent, predict what he will do. But the man has no thought for the slater, nor the slater for him; they seem to belong to two worlds completely foreign to one another. Nevertheless the slater drops a tile which kills the man, and we should have no hesitation in saying that this was chance.

Our frailty does not permit us to take in the whole universe, but forces us to cut it up in slices. We attempt to make this as little artificial as possible, and yet it happens, from time to time, that two of these slices react upon each other, and then the effects of this mutual action appear to us to be due to chance.

Is this a third way of conceiving of chance? Not always; in fact, in the majority of cases, we come back to the first or second. Each time that two worlds, generally foreign to one another, thus come to act upon each other, the laws of this reaction cannot fail to be very complex, and moreover a very small change in the initial conditions of the two worlds would have been enough to prevent the reaction from taking place. How very little it would have taken to make the man pass a moment later, or the slater drop his tile a moment earlier!

VI.

Nothing that has been said so far explains why chance is obedient to laws. Is the fact that the causes are small, or that they are complex, sufficient to enable us to predict, if not what the effects will be *in each case*, at least what they will be *on the average?* In order to answer this question, it will be best to return to some of the examples quoted above.

I will begin with that of roulette. I said that the point where the needle stops will depend on the initial impulse given it. What is the probability that this impulse will be of any particular strength? I do not know, but it is difficult not to admit that this probability is represented by a continuous analytical function. The probability that the impulse will be comprised between α and $\alpha+\epsilon$ will, then, clearly be equal to the probability that it will be comprised between $\alpha+\epsilon$ and $\alpha+2\epsilon$, *provided that ϵ is very small.* This is a property common to all analytical functions. Small variations of the function are proportional to small variations of the variable.

But we have assumed that a very small variation in the impulse is sufficient to change the colour of the section opposite which the needle finally stops. From α to $\alpha+\epsilon$ is red, from $\alpha+\epsilon$ to $\alpha+2\epsilon$ is black. The probability of each red section is accordingly the same as that of the succeeding black section, and consequently the total probability of red is equal to the total probability of black.

The datum in the case is the analytical function which represents the probability of a particular initial impulse. But the theorem remains true, whatever this datum may be, because it depends on a property common to all analytical functions. From this it results finally that we have no longer any need of the datum.

What has just been said of the case of roulette applies also to the example of the minor planets. The Zodiac may be regarded as an immense roulette board on which the Creator has thrown a very great number of small balls, to which he has imparted different initial impulses, varying, however, according to some sort of law. Their actual distribution is uniform and independent of that law, for the same reason as in the preceding case. Thus we see why phenomena obey the laws of chance when small differences in the causes are sufficient to produce great differences in the effects. The probabilities of these small differences can then be regarded as proportional to the differences themselves, just because these differences are small, and small increases of a continuous function are proportional to those of the variable.

Let us pass to a totally different example, in which the complexity of the causes is the principal factor. I imagine a card-player shuffling a pack of cards. At each shuffle he changes the order of the cards, and he may change it in various ways. Let us take three cards only in order to simplify the explanation. The cards which, before the shuffle, occupied the positions 1 2 3 respectively may, after the shuffle, occupy the positions

1 2 3, 2 3 1, 3 1 2, 3 2 1, 1 3 2, 2 1 3.

Each of these six hypotheses is possible, and their probabilities are respectively

$$p_1, p_2, p_3, p_4, p_5, p_6.$$

The sum of these six numbers is equal to 1, but that is all we know about them. The six probabilities naturally depend upon the player's habits, which we do not know.

At the second shuffle the process is repeated, and under the same conditions. I mean, for instance, that p_4 always represents the probability that the three cards which occupied the positions 1 2 3 after the nth shuffle and before the $n+1^{th}$, will occupy the positions 3 2 1 after the $n+1^{th}$ shuffle. And this remains true, whatever the number n may be, since the player's habits and his method of shuffling remain the same.

But if the number of shuffles is very large, the cards which occupied the positions 1 2 3 before the first shuffle may, after the last shuffle, occupy the positions

$$123,\ 231,\ 312,\ 321,\ 132,\ 213,$$

and the probability of each of these six hypotheses is clearly the same and equal to $\frac{1}{6}$; and this is true whatever be the numbers $p_1 \ldots p_6$, which we do not know. The great number of shuffles, that is to say, the complexity of the causes, has produced uniformity.

This would apply without change if there were more than three cards, but even with three the demonstration would be complicated, so I will content myself with giving it for two cards only. We have now only two hypotheses

$$12,\ 21,$$

with the probabilities p_1 and $p_2 = 1 - p_1$. Assume that there are n shuffles, and that I win a shilling if the cards are finally in the initial order, and that I lose one if they are finally reversed. Then my mathematical expectation will be

$$(p_1 - p_2)^n$$

The difference $p_1 - p_2$ is certainly smaller than 1, so that if n is very large, the value of my expectation will be nothing, and we do not require to know p_1 and p_2 to know that the game is fair.

Nevertheless there would be an exception if one of the numbers p_1 and p_2 was equal to 1 and the other to nothing. *It would then hold good no longer, because our original hypotheses would be too simple.*

What we have just seen applies not only to the

mixing of cards, but to all mixing, to that of powders and liquids, and even to that of the gaseous molecules in the kinetic theory of gases. To return to this theory, let us imagine for a moment a gas whose molecules cannot collide mutually, but can be deviated by collisions with the sides of the vessel in which the gas is enclosed. If the form of the vessel is sufficiently complicated, it will not be long before the distribution of the molecules and that of their velocities become uniform. This will not happen if the vessel is spherical, or if it has the form of a rectangular parallelepiped. And why not? Because in the former case the distance of any particular trajectory from the centre remains constant, and in the latter case we have the absolute value of the angle of each trajectory with the sides of the parallelepiped.

Thus we see what we must understand by conditions that are *too simple*. They are conditions which preserve something of the original state as an invariable. Are the differential equations of the problem too simple to enable us to apply the laws of chance? This question appears at first sight devoid of any precise meaning, but we know now what it means. They are too simple if something is preserved, if they admit a uniform integral. If something of the initial conditions remains unchanged, it is clear that the final situation can no longer be independent of the initial situation.

We come, lastly, to the theory of errors. We are ignorant of what accidental errors are due to, and it is just because of this ignorance that we know they will obey Gauss's law. Such is the paradox. It is explained in somewhat the same way as the preceding cases. We only need to know one thing—that the errors are very numerous, that they are very small, and that each of them can be equally well negative or positive. What is the curve of probability of each of them? We do not know, but only assume that it is symmetrical. We can then show that the resultant error will follow Gauss's law, and this resultant law is independent of the particular laws which we do not know. Here again the simplicity of the result actually owes its existence to the complication of the data.

VII.

But we have not come to the end of paradoxes. I recalled just above Flammarion's fiction of the man who travels faster than light, for whom time has its sign changed. I said that for him all phenomena would seem to be due to chance. This is true from a certain point of view, and yet, at any given moment, all these phenomena would not be distributed in conformity with the laws of chance, since they would be just as they are for us, who, seeing them unfolded harmoniously and not emerging from a primitive chaos, do not look upon them as governed by chance.

What does this mean? For Flammarion's imaginary Lumen, small causes seem to produce great effects; why, then, do things not happen as they do for us when we think we see great effects due to small causes? Is not the same reasoning applicable to his case?

Let us return to this reasoning. When small differences in the causes produce great differences in the effects, why are the effects distributed according to the laws of chance? Suppose a difference of an inch in the cause produces a difference of a mile in the effect. If I am to win in case the effect corresponds with a mile bearing an even number, my probability of winning will be $\frac{1}{2}$. Why is this? Because, in order that it should be so, the cause must correspond with an inch bearing an even number. Now, according to all appearance, the probability that the cause will vary between certain limits is proportional to the distance of those limits, provided that distance is very small. If this hypothesis be not admitted, there would no longer be any means of representing the probability by a continuous function.

Now what will happen when great causes produce small effects? This is the case in which we shall not attribute the phenomenon to chance, and in which Lumen, on the contrary, would attribute it to chance. A difference of a mile in the cause corresponds to a difference of an inch in the effect. Will the probability that the cause will be comprised between two limits n miles apart still be proportional to n? We have no reason to suppose it, since this distance of n miles is great. But the probability that the effect will be comprised between two limits n inches apart will be precisely the same, and accordingly it will not be proportional to n, and that notwithstanding the fact that this distance of n inches is small. There is, then, no means of representing the law of probability of the effects by a continuous curve. I do not mean to say that the curve may not remain continuous in the *analytical* sense of the word. To *infinitely small* variations of the abscissa there will correspond infinitely small variations of the ordinate. But *practically* it would not be continuous, since to *very small* variations of the abscissa there would not correspond very small variations of the ordinate. It would become impossible to trace the curve with an ordinary pencil: that is what I mean.

What conclusion are we then to draw? Lumen has no right to say that the probability of the cause (that of *his* cause, which is our effect) must necessarily be represented by a continuous function. But if that be so, why have we the right? It is because that state of unstable equilibrium that I spoke of just now as initial, is itself only the termination of a long anterior history. In the course of this history complex causes have been at work, and they have been at work for a long time. They have contributed to bring about the mixture of the elements, and they have tended to make everything uniform, at least in a small space. They have rounded off the corners, levelled the mountains, and filled up the valleys. However capricious and irregular the original curve they have been given, they have worked so much to regularize it that they will finally give us a continuous curve, and that is why we can quite confidently admit its continuity.

Lumen would not have the same reasons for drawing this conclusion. For him complex causes would not appear as agents of regularity and of levelling; on the contrary, they would only create differentiation and inequality. He would see a more and more varied world emerge from a sort of primitive chaos. The changes he would observe would be for him unforeseen and impossible to foresee. They would seem to him due to some caprice, but that caprice would not be at all the same as our chance, since it would not be amenable to any law, while our chance has its own laws. All these points would require a much longer development, which would help us perhaps to a better comprehension of the irreversibility of the universe.

VIII.

We have attempted to define chance, and it would be well now to ask ourselves a question. Has chance, thus defined so far as it can be, an objective character?

We may well ask it. I have spoken of very small or very complex causes, but may not what is very small for one be great for another, and may not what seems very complex to one appear simple to another? I have already given a partial answer, since I stated above most precisely the case in which differential equations become too simple for the laws of chance to remain applicable. But it would be well to examine the thing somewhat more closely, for there are still other points of view we may take.

What is the meaning of the word small? To understand it, we have only to refer to what has been said above. A difference is very small, an interval is small, when within the limits of that interval the probability remains appreciably constant. Why can that probability be regarded as constant in a small interval? It is because we admit that the law of probability is represented by a continuous curve, not only continuous in the analytical sense of the word, but *practically* continuous, as I explained above. This means not only that it will present no absolute hiatus, but also that it will have no projections or depressions too acute or too much accentuated.

What gives us the right to make this hypothesis?

As I said above, it is because, from the beginning of the ages, there are complex causes that never cease to operate in the same direction, which cause the world to tend constantly towards uniformity without the possibility of ever going back. It is these causes which, little by little, have levelled the projections and filled up the depressions, and it is for this reason that our curves of probability present none but gentle undulations. In millions and millions of centuries we shall have progressed another step towards uniformity, and these undulations will be ten times more gentle still. The radius of mean curvature of our curve will have become ten times longer. And then a length that to-day does not seem to us very small, because an arc of such a length cannot be regarded as rectilineal, will at that period be properly qualified as very small, since the curvature will have become ten times less, and an arc of such a length will not differ appreciably from a straight line.

Thus the word very small remains relative, but it is not relative to this man or that, it is relative to the actual state of the world. It will change its meaning when the world becomes more uniform and all things are still more mixed. But then, no doubt, men will no longer be able to live, but will have to make way for other beings, shall I say much smaller or much larger? So that our criterion, remaining true for all men, retains an objective meaning.

And, further, what is the meaning of the word very complex? I have already given one solution, that which I referred to again at the beginning of this section; but there are others. Complex causes, I have said, produce a more and more intimate mixture, but how long will it be before this mixture satisfies us? When shall we have accumulated enough complications? When will the cards be sufficiently shuffled? If we mix two powders, one blue and the other white, there comes a time when the colour of the mixture appears uniform. This is on account of the infirmity of our senses; it would be uniform for the long-sighted, obliged to look at it from a distance, when it would not yet be so for the short-sighted. Even when it had become uniform for all sights, we could still set back the limit by employing instruments. There is no possibility that any man will ever distinguish the infinite variety that is hidden under the uniform appearance of a gas, if the kinetic theory is true. Nevertheless, if we adopt Gouy's ideas on the Brownian movement, does not the microscope seem to be on the point of showing us something analogous?

This new criterion is thus relative like the first, and if it preserves an objective character, it is because all men have about the same senses, the power of their instruments is limited, and, moreover, they only make use of them occasionally.

IX.

It is the same in the moral sciences, and particularly in history. The historian is obliged to make a selection of the events in the period he is studying, and he only recounts those that seem to him the most important. Thus he contents himself with relating the most considerable events of the 16th century, for instance, and similarly the most remarkable facts of the 17th century. If the former are sufficient to explain the latter, we say that these latter conform to the laws of history. But if a great event of the 17th century owes its cause to a small fact of the 16th century that no history reports and that every one has neglected, then we say that this event is due to chance, and so the word has the same sense as in the physical sciences; it means that small causes have produced great effects.

The greatest chance is the birth of a great man. It is only by chance that the meeting occurs of two genital cells of different sex that contain precisely, each on its side, the mysterious elements whose mutual reaction is destined to produce genius. It will be readily admitted that these elements must be rare, and that their meeting is still rarer. How little it would have taken to make the spermatozoid which carried them deviate from its course. It would have been enough to deflect it a hundredth part of a inch, and Napoleon would not have been born and the destinies of a continent would have been changed. No example can give a better comprehension of the true character of chance.

One word more about the paradoxes to which the application of the calculation of probabilities to the moral sciences has given rise. It has been demonstrated that no parliament would ever contain a single member of the opposition, or at least that such an event would be so improbable that it would be quite safe to bet against it, and to bet a million to one. Condorcet attempted to calculate how many jurymen it would require to make a miscarriage of justice practically impossible. If we used the results of this calculation, we should certainly be exposed to the same disillusionment as by betting on the strength of the calculation that the opposition would never have a single representative.

The laws of chance do not apply to these questions. If justice does not always decide on good grounds, it does not make so much use as is generally supposed of Bridoye's method. This is perhaps unfortunate, since, if it did, Condorcet's method would protect us against miscarriages.

What does this mean? We are tempted to attribute facts of this nature to chance because their causes are obscure, but this is not true chance. The causes are unknown to us, it is true, and they are even complex; but they are not sufficiently complex, since they preserve something, and we have seen that this is the distinguishing mark of "too simple" causes. When men are brought together, they no longer decide by chance and independently of each other, but react upon one another. Many causes come into action, they trouble the men and draw them this way and that, but there is one thing they cannot destroy, the habits they have of Panurge's sheep. And it is this that is preserved.

X.

The application of the calculation of probabilities to the exact sciences also involves many difficulties. Why are the decimals of a table of logarithms or of the number π distributed in accordance with the laws of chance? I have elsewhere studied the question in regard to logarithms, and there it is easy. It is clear that a small difference in the argument will give a small difference in the logarithm, but a great difference in the sixth decimal of the logarithm. We still find the same criterion.

But as regards the number π the question presents more difficulties, and for the moment I have no satisfactory explanation to give.

There are many other questions that might be raised, if I wished to attack them before answering the one I have more especially set myself. When we arrive at a simple result, when, for instance, we find a round number, we say that such a result cannot be due to chance, and we seek for a non-fortuitous cause to explain it. And in fact there is only a very slight likelihood that, out of 10,000 numbers, chance will give us a round number, the number 10,000 for instance; there is only one chance in 10,000. But neither is there more than one chance in 10,000 that it will give us any other particular number, and yet this result does not astonish us, and we feel no hesitation about attributing it to chance, and that merely because it is less striking.

Is this a simple illusion on our part, or are there cases in which this view is legitimate? We must hope so, for otherwise all science would be impossible. When we wish to check a hypothesis, what do we do? We cannot verify all its consequences, since they are infinite in number. We content ourselves with verifying a few, and, if we succeed, we declare that the hypothesis is confirmed, for so much success could not be due to chance. It is always at bottom the same reasoning.

I cannot justify it here completely, it would take me too long, but I can say at least this. We find ourselves faced by two hypotheses, either a simple cause or else that assemblage of complex causes we call chance. We find it natural to admit that the former must produce a simple result, and then, if we arrive at this simple result, the round number for instance, it appears to us more reasonable to attribute it to the simple cause, which was almost certain to give it us, than to chance, which could only give it us once in 10,000 times. It will not be the same if we arrive at a result that is not simple. It is true that chance also will not give it more than once in 10,000 times, but the simple cause has no greater chance of producing it.

BOOK II.

MATHEMATICAL REASONING.

I.

THE RELATIVITY OF SPACE.

I.

IT is impossible to picture empty space. All our efforts to imagine pure space from which the changing images of material objects are excluded can only result in a representation in which highly-coloured surfaces, for instance, are replaced by lines of slight colouration, and if we continued in this direction to the end, everything would disappear and end in nothing. Hence arises the irreducible relativity of space.

Whoever speaks of absolute space uses a word devoid of meaning. This is a truth that has been long proclaimed by all who have reflected on the question, but one which we are too often inclined to forget.

If I am at a definite point in Paris, at the Place du Panthéon, for instance, and I say, "I will come back *here* to-morrow;" if I am asked, "Do you mean that you will come back to the same point in space?" I should be tempted to answer yes. Yet I should be wrong, since between now and to-morrow the earth will have moved, carrying with it the Place du Panthéon, which will have travelled more than a million miles. And if I wished to speak more accurately, I should gain nothing, since this million of miles has been covered by our globe in its motion in relation to the sun, and the sun in its turn moves in relation to the Milky Way, and the Milky Way itself is no doubt in motion without our being able to recognize its velocity. So that we are, and shall always be, completely ignorant how far the Place du Panthéon moves in a day. In fact, what I meant to say was, "To-morrow I shall see once more the dome and pediment of the Panthéon," and if there was no Panthéon my sentence would have no meaning and space would disappear.

This is one of the most commonplace forms of the principle of the relativity of space, but there is another on which Delbeuf has laid particular stress. Suppose that in one night all the dimensions of the universe became a thousand times larger. The world will remain *similar* to itself, if we give the word *similitude* the meaning it has in the third book of Euclid. Only, what was formerly a metre long will now measure a kilometre, and what was a millimetre long will become a metre. The bed in which I went to sleep and my body itself will have grown in the same proportion. When I wake in the morning what will be my feeling in face of such an astonishing transformation? Well, I shall not notice anything at all. The most exact measures will be incapable of revealing anything of this tremendous change, since the yard-measures I shall use will have varied in exactly the same proportions as the objects I shall attempt to measure. In reality the change only exists for those who argue as if space were absolute. If I have argued for a moment as they do, it was only in order to make it clearer that their view implies a contradiction. In reality it would be better to say that as space is relative, nothing at all has happened, and that it is for that reason that we have noticed nothing.

Have we any right, therefore, to say that we know the distance between two points? No, since that distance could undergo enormous variations without our being able to perceive it, provided other distances varied in the same proportions. We saw just now that when I say I shall be here to-morrow, that does not mean that to-morrow I shall be at the point in space where I am to-day, but that to-morrow I shall be at the same distance from the Panthéon as I am to-day. And already this statement is not sufficient, and I ought to say that to-morrow and to-day my distance from the Panthéon will be equal to the same number of times the length of my body.

But that is not all. I imagined the dimensions of the world changing, but at least the world remaining always similar to itself. We can go much further than that, and one of the most surprising theories of modern physicists will furnish the occasion. According to a hypothesis of Lorentz and Fitzgerald,* all bodies carried forward in the earth's motion undergo a deformation. This deformation is, in truth, very slight, since all dimensions parallel with the earth's motion are diminished by a hundred-millionth, while dimensions perpendicular to this motion are not altered. But it matters little that it is slight; it is enough that it should exist for the conclusion I am soon going to draw from it. Besides, though I said that it is slight, I really know nothing about it. I have myself fallen a victim to the tenacious illusion that

* *Vide infra*, Book III. Chap. ii.

makes us believe that we think of an absolute space. I was thinking of the earth's motion on its elliptical orbit round the sun, and I allowed 18 miles a second for its velocity. But its true velocity (I mean this time, not its absolute velocity, which has no sense, but its velocity in relation to the ether), this I do not know and have no means of knowing. It is, perhaps, 10 or 100 times as high, and then the deformation will be 100 or 10,000 times as great.

It is evident that we cannot demonstrate this deformation. Take a cube with sides a yard long. It is deformed on account of the earth's velocity; one of its sides, that parallel with the motion, becomes smaller, the others do not vary. If I wish to assure myself of this with the help of a yard-measure, I shall measure first one of the sides perpendicular to the motion, and satisfy myself that my measure fits this side exactly; and indeed neither one nor other of these lengths is altered, since they are both perpendicular to the motion. I then wish to measure the other side, that parallel with the motion; for this purpose I change the position of my measure, and turn it so as to apply it to this side. But the yard-measure, having changed its direction and having become parallel with the motion, has in its turn undergone the deformation, so that, though the side is no longer a yard long, it will still fit it exactly, and I shall be aware of nothing.

What, then, I shall be asked, is the use of the hypothesis of Lorentz and Fitzgerald if no experiment can enable us to verify it? The fact is that my statement has been incomplete. I have only spoken of measurements that can be made with a yard-measure, but we can also measure a distance by the time that light takes to traverse it, on condition that we admit that the velocity of light is constant, and independent of its direction. Lorentz could have accounted for the facts by supposing that the velocity of light is greater in the direction of the earth's motion than in the perpendicular direction. He preferred to admit that the velocity is the same in the two directions, but that bodies are smaller in the former than in the latter. If the surfaces of the waves of light had undergone the same deformations as material bodies, we should never have perceived the Lorentz-Fitzgerald deformation.

In the one case as in the other, there can be no question of absolute magnitude, but of the measurement of that magnitude by means of some instrument. This instrument may be a yard-measure or the path traversed by light. It is only the relation of the magnitude to the instrument that we measure, and if this relation is altered, we have no means of knowing whether it is the magnitude or the instrument that has changed.

But what I wish to make clear is, that in this deformation the world has not remained similar to itself. Squares have become rectangles or parallelograms, circles ellipses, and spheres ellipsoids. And yet we have no means of knowing whether this deformation is real.

It is clear that we might go much further. Instead of the Lorentz-Fitzgerald deformation, with its extremely simple laws, we might imagine a deformation of any kind whatever; bodies might be deformed in accordance with any laws, as complicated as we liked, and we should not perceive it, provided all bodies without exception were deformed in accordance with the same laws. When I say all bodies without exception, I include, of course, our own bodies and the rays of light emanating from the different objects.

If we look at the world in one of those mirrors of complicated form which deform objects in an odd way, the mutual relations of the different parts of the world are not altered; if, in fact, two real objects touch, their images likewise appear to touch. In truth, when we look in such a mirror we readily perceive the deformation, but it is because the real world exists beside its deformed image. And even if this real world were hidden from us, there is something which cannot be hidden, and that is ourselves. We cannot help seeing, or at least feeling, our body and our members which have not been deformed, and continue to act as measuring instruments. But if we imagine our body itself deformed, and in the same way as if it were seen in the mirror, these measuring instruments will fail us in their turn, and the deformation will no longer be able to be ascertained.

Imagine, in the same way, two universes which are the image one of the other. With each object P in the universe A, there corresponds, in the universe B, an object P^1 which is its image. The co-ordinates of this image P^1 are determinate functions of those of the object P; moreover, these functions may be of any kind whatever—I assume only that they are chosen once for all. Between the position of P and that of P^1 there is a constant relation; it matters little what that relation may be, it is enough that it should be constant.

Well, these two universes will be indistinguishable.

I mean to say that the former will be for its inhabitants what the second is for its own. This would be true so long as the two universes remained foreign to one another. Suppose we are inhabitants of the universe A; we have constructed our science and particularly our geometry. During this time the inhabitants of the universe B have constructed a science, and as their world is the image of ours, their geometry will also be the image of ours, or, more accurately, it will be the same. But if one day a window were to open for us upon the universe B, we should feel contempt for them, and we should say, "These wretched people imagine that they have made a geometry, but what they so name is only a grotesque image of ours; their straight lines are all twisted, their circles are hunchbacked, and their spheres have capricious inequalities." We should have no suspicion that they were saying the same of us, and that no one will ever know which is right.

We see in how large a sense we must understand the relativity of space. Space is in reality amorphous, and it is only the things that are in it that give it a form. What are we to think, then, of that direct intuition we have of a straight line or of distance? We have so little the intuition of distance in itself that, in a single night, as we have said, a distance could become a thousand times greater without our being able to perceive it, if all other distances had undergone the same alteration. And in a night the universe B might even be substituted for the universe A without our having any means of knowing it, and then the straight lines of yesterday would have ceased to be straight, and we should not be aware of anything.

One part of space is not by itself and in the absolute sense of the word equal to another part of space, for if it is so for us, it will not be so for the inhabitants of the universe B, and they have precisely as much right to reject our opinion as we have to condemn theirs.

I have shown elsewhere what are the consequences of these facts from the point of view of the idea that we should construct non-Euclidian and other analogous geometries. I do not wish to return to this, and I will take a somewhat different point of view.

II.

If this intuition of distance, of direction, of the straight line, if, in a word, this direct intuition of space does not exist, whence comes it that we imagine we have it? If this is only an illusion, whence comes it that the illusion is so tenacious? This is what we must examine. There is no direct intuition of magnitude, as we have said, and we can only arrive at the relation of the magnitude to our measuring instruments. Accordingly we could not have constructed space if we had not had an instrument for measuring it. Well, that instrument to which we refer everything, which we use instinctively, is our own body. It is in reference to our own body that we locate exterior objects, and the only special relations of these objects that we can picture to ourselves are their relations with our body. It is our body that serves us, so to speak, as a system of axes of co-ordinates.

For instance, at a moment α the presence of an object A is revealed to me by the sense of sight; at another moment β the presence of another object

B is revealed by another sense, that, for instance, of hearing or of touch. I judge that this object B occupies the same place as the object A. What does this mean? To begin with, it does not imply that these two objects occupy, at two different moments, the same point in an absolute space, which, even if it existed, would escape our knowledge, since between the moments α and β the solar system has been displaced and we cannot know what this displacement is. It means that these two objects occupy the same relative position in reference to our body.

But what is meant even by this? The impressions that have come to us from these objects have followed absolutely different paths—the optic nerve for the object A, and the acoustic nerve for the object B; they have nothing in common from the qualitative point of view. The representations we can form of these two objects are absolutely heterogeneous and irreducible one to the other. Only I know that, in order to reach the object A, I have only to extend my right arm in a certain way; even though I refrain from doing it, I represent to myself the muscular and other analogous sensations which accompany that extension, and that representation is associated with that of the object A.

Now I know equally that I can reach the object B by extending my right arm in the same way, an extension accompanied by the same train of muscular sensations. And I mean nothing else but this when I say that these two objects occupy the same position.

I know also that I could have reached the object A by another appropriate movement of the left arm, and I represent to myself the muscular sensations that would have accompanied the movement. And by the same movement of the left arm, accompanied by the same sensations, I could equally have reached the object B.

And this is very important, since it is in this way that I could defend myself against the dangers with which the object A or the object B might threaten me. With each of the blows that may strike us, nature has associated one or several parries which enable us to protect ourselves against them. The same parry may answer to several blows. It is thus, for instance, that the same movement of the right arm would have enabled us to defend ourselves at the moment α against the object A, and at the moment β against the object B. Similarly, the same blow may be parried in several ways, and we have said, for instance, that we could reach the object A equally well either by a certain movement of the right arm, or by a certain movement of the left.

All these parries have nothing in common with one another, except that they enable us to avoid the same blow, and it is that, and nothing but that, we mean when we say that they are movements ending in the same point in space. Similarly, these objects, of which we say that they occupy the same point in space, have nothing in common, except that the same parry can enable us to defend ourselves against them.

Or, if we prefer it, let us imagine innumerable telegraph wires, some centripetal and others centrifugal. The centripetal wires warn us of accidents that occur outside, the centrifugal wires have to provide the remedy. Connexions are established in such a way that when one of the centripetal wires is traversed by a current, this current acts on a central exchange, and so excites a current in one of the centrifugal wires, and matters are so arranged that several centripetal wires can act on the same centrifugal wire, if the same remedy is applicable to several evils, and that one centripetal wire can disturb several centrifugal wires, either simultaneously or one in default of the other, every time that the same evil can be cured by several remedies.

It is this complex system of associations, it is this distribution board, so to speak, that is our whole geometry, or, if you will, all that is distinctive in our geometry. What we call our intuition of a straight line or of distance is the consciousness we have of these associations and of their imperious character.

Whence this imperious character itself comes, it is easy to understand. The older an association is, the more indestructible it will appear to us. But these associations are not, for the most part, conquests made by the individual, since we see traces of them in the newly-born infant; they are conquests made by the race. The more necessary these conquests were, the more quickly they must have been brought about by natural selection.

On this account those we have been speaking of must have been among the earliest, since without them the defence of the organism would have been impossible. As soon as the cells were no longer merely in juxtaposition, as soon as they were called upon to give mutual assistance to each other, some such mechanism as we have been describing must necessarily have been organized in order that the assistance should meet the danger without miscarrying.

When a frog's head has been cut off, and a drop of acid is placed at some point on its skin, it tries to rub off the acid with the nearest foot; and if that foot is cut off, it removes it with the other foot. Here we have, clearly, that double parry I spoke of just now, making it possible to oppose an evil by a second remedy if the first fails. It is this multiplicity of parries, and the resulting co-ordination, that is space.

We see to what depths of unconsciousness we have to descend to find the first traces of these spacial associations, since the lowest parts of the nervous system alone come into play. Once we have realized this, how can we be astonished at the resistance we oppose to any attempt to dissociate what has been so long associated? Now, it is this very resistance that we call the evidence of the truths of geometry. This evidence is nothing else than the repugnance we feel at breaking with very old habits with which we have always got on very well.

III.

The space thus created is only a small space that does not extend beyond what my arm can reach, and the intervention of memory is necessary to set back its limits. There are points that will always remain out of my reach, whatever effort I may make to stretch out my hand to them. If I were attached to the ground, like a sea-polype, for instance, which can only extend its tentacles, all these points would be outside space, since the sensations we might experience from the action of bodies placed there would not be associated with the idea of any movement enabling us to reach them, or with any appropriate parry. These sensations would not seem to us to have any spacial character, and we should not attempt to locate them.

But we are not fixed to the ground like the inferior animals. If the enemy is too far off, we can advance upon him first and extend our hand when we are near enough. This is still a parry, but a long-distance parry. Moreover, it is a complex parry, and into the representation we make of it there enter the representation of the muscular sensations caused by the movement of the legs, that of the muscular sensations caused by the final movement of the arm, that of the sensations of the semi-circular canals, etc. Besides, we have to make a representation, not of a complexus of simultaneous sensations, but of a complexus of successive sensations, following one another in a determined order, and it is for this reason that I said just now that the intervention of memory is necessary.

We must further observe that, to reach the same point, I can approach nearer the object to be attained, in order not to have to extend my hand so far. And how much more might be said? It is not one only, but a thousand parries I can oppose to the same danger. All these parries are formed of sensations that may have nothing in common, and yet we regard them as defining the same point in space, because they can answer to the same danger and are one and all of them associated with the notion of that danger. It is the possibility of parrying the same blow which makes the unity of these different parries, just as it is the possibility of being parried in the same way which makes the unity of the blows of such different kinds that can threaten us from the same point in space. It is this double unity that makes the individuality of each point in space, and in the notion of such a point there is nothing else but this.

The space I pictured in the preceding section, which I might call *restricted space*, was referred to axes of co-ordinates attached to my body. These axes were fixed, since my body did not move, and it was only my limbs that changed their position. What are the axes to which the *extended space* is naturally referred—that is to say, the new space I have just defined? We define a point by the succession of movements we require to make to reach it, starting from a certain initial position of the body. The axes are accordingly attached to this initial position of the body.

But the position I call initial may be arbitrarily chosen from among all the positions my body has successively occupied. If a more or less unconscious memory of these successive positions is necessary for the genesis of the notion of space, this memory can go back more or less into the past. Hence results a certain indeterminateness in the very definition of space, and it is precisely this indeterminateness which constitutes its relativity.

Absolute space exists no longer; there is only space relative to a certain initial position of the body. For a conscious being, fixed to the ground like the inferior animals, who would consequently only know restricted space, space would still be relative, since it would be referred to his body, but this being would not be conscious of the relativity, because the axes to which he referred this restricted space would not change. No doubt the rock to which he was chained would not be motionless, since it would be involved in the motion of our planet; for us, consequently, these axes would change every moment, but for him they would not change. We have the faculty of referring our extended space at one time to the position A of our body considered as initial, at another to the position B which it occupied some moments later, which we are free to consider in its turn as initial, and, accordingly, we make unconscious changes in the co-ordinates every moment. This faculty would fail our imaginary being, and, through not having travelled, he would think space absolute. Every moment his system of axes would be imposed on him; this system might change to any extent in reality, for him it would be always the same, since it would always be the *unique* system. It is not the same for us who possess, each moment, several systems between which we can choose at will, and on condition of going back by memory more or less into the past.

That is not all, for the restricted space would not be homogeneous. The different points of this space could not be regarded as equivalent, since some could only be reached at the cost of the greatest efforts, while others could be reached with ease. On the contrary, our extended space appears to us homogeneous, and we say that all its points are equivalent. What does this mean?

If we start from a certain position A, we can, starting from that position, effect certain movements M, characterized by a certain complexus of muscular sensations. But, starting from another position B, we can execute movements M^1 which will be characterized by the same muscular sensations. Then let a be the situation of a certain point in the body, the tip of the forefinger of the right hand, for instance, in the initial position A, and let b be the position of this same forefinger when, starting from that position A, we have executed the movements M. Then let a^1 be the situation of the forefinger in the position B, and b^1 its situation when, starting from the position B, we have executed the movements M^1.

Well, I am in the habit of saying that the points a and b are, in relation to each other, as the points a^1 and b^1, and that means simply that the two series of movements M and M^1 are accompanied by the same muscular sensations. And as I am conscious that, in passing from the position A to the position B, my body has remained capable of the same movements, I know that there is a point in space which is to the point a^1 what some point b is to the point a, so that the two points a and a^1 are equivalent. It is this that is called the homogeneity of space, and at the same time it is for this reason that space is relative, since its properties remain the same whether they are referred to the axes A or to the axes B. So that the relativity of space and its homogeneity are one and the same thing.

Now, if I wish to pass to the great space, which is no longer to serve for my individual use only, but in which I can lodge the universe, I shall arrive at it by an act of imagination. I shall imagine what a giant would experience who could reach the planets in a few steps, or, if we prefer, what I should feel myself in presence of a world in miniature, in which these planets would be replaced by little balls, while on one of these little balls there would move a Lilliputian that I should call myself. But this act of imagination would be impossible for me if I had not previously constructed my restricted space and my extended space for my personal use.

IV.

Now we come to the question why all these spaces have three dimensions. Let us refer to the "distribution board" spoken of above. We have, on the one side, a list of the different possible dangers—let us designate them as A1, A2, etc.—and, on the other side, the list of the different remedies, which I will call in the same way B1, B2, etc. Then we have connexions between the contact studs of the first list and those of the second in such a way that when, for instance, the alarm for danger A3 works, it sets in motion or may set in motion the relay corresponding to the parry B4.

As I spoke above of centripetal or centrifugal wires, I am afraid that all I have said may be taken, not as a simple comparison, but as a description of the nervous system. Such is not my thought, and that for several reasons. Firstly, I should not presume to pronounce an opinion on the structure of the nervous system which I do not know, while those who have studied it only do so with circumspection. Secondly, because, in spite of my incompetence, I fully realize that this scheme would be far too simple. And lastly, because, on my list of parries, there appear some that are very complex, which may even, in the case of extended space, as we have seen above, consist of several steps followed by a movement of the arm. It is not a question, then, of physical connexion between two real conductors, but of psychological association between two series of sensations.

If A1 and A2, for instance, are both of them associated with the parry B1, and if A1 is similarly associated with B2, it will generally be the case that A2 and B2 will also be associated. If this fundamental law were not generally true, there would only be an immense confusion, and there would be nothing that could bear any resemblance to a conception of space or to a geometry. How, indeed, have we defined a point in space? We defined it in two ways: on the one hand, it is the whole of the alarms A which are in connexion with the same parry B; on the other, it is the whole of the parries B which are in connexion with the same alarm A. If our law were not true, we should be obliged to say that A1 and A2 correspond with the same point, since they are both in connexion with B1; but we should be equally obliged to say that they do not correspond with the same point, since A1 would be in connexion with B2, and this would not be true of A2—which would be a contradiction.

But from another aspect, if the law were rigorously and invariably true, space would be quite different from what it is. We should have well-defined categories, among which would be apportioned the alarms A on the one side and the parries B on the other. These categories would be exceedingly numerous, but they would be entirely separated one from the other. Space would be formed of points, very numerous but discrete; it would be *discontinuous*. There would be no reason for arranging these points in one order rather than another, nor, consequently, for attributing three dimensions to space.

But this is not the case. May I be permitted for a moment to use the language of those who know geometry already? It is necessary that I should do so, since it is the language best understood by those to whom I wish to make myself clear. When I wish to parry the blow, I try to reach the point whence the blow comes, but it is enough if I come fairly near it. Then the parry B1 may answer to A1, and to A2 if the point which corresponds with B1 is sufficiently close both to that which corresponds with A1 and to that which corresponds with A2. But it may happen that the point which corresponds with another parry B2 is near enough to the point corresponding with A1, and not near enough to the point corresponding with A2. And so the parry B2 may answer to A1 and not be able to answer to A2.

For those who do not yet know geometry, this may be translated simply by a modification of the law enunciated above. Then what happens is as follows. Two parries, B1 and B2, are associated with one alarm A1, and with a very great number of alarms that we will place in the same category as A1, and make to correspond with the same point in space. But we may find alarms A2 which are associated with B2 and not with B1, but on the other hand are associated with B3, which are not with A1, and so on in succession, so that we may write the sequence

B1, A1, B2, A2, B3, A3, B4, A4,

in which each term is associated with the succeeding

and preceding terms, but not with those that are several places removed.

It is unnecessary to add that each of the terms of these sequences is not isolated, but forms part of a very numerous category of other alarms or other parries which has the same connexions as it, and may be regarded as belonging to the same point in space. Thus the fundamental law, though admitting of exceptions, remains almost always true. Only, in consequence of these exceptions, these categories, instead of being entirely separate, partially encroach upon each other and mutually overlap to a certain extent, so that space becomes continuous.

Furthermore, the order in which these categories must be arranged is no longer arbitrary, and a reference to the preceding sequence will make it clear that B2 must be placed between A1 and A2, and, consequently, between B1 and B3, and that it could not be placed, for instance, between B3 and B4.

Accordingly there is an order in which our categories range themselves naturally which corresponds with the points in space, and experience teaches us that this order presents itself in the form of a three-circuit distribution board, and it is for this reason that space has three dimensions.

V.

Thus the characteristic property of space, that of having three dimensions, is only a property of our distribution board, a property residing, so to speak, in the human intelligence. The destruction of some of these connexions, that is to say, of these associations of ideas, would be sufficient to give us a different distribution board, and that might be enough to endow space with a fourth dimension.

Some people will be astonished at such a result. The exterior world, they think, must surely count for something. If the number of dimensions comes from the way in which we are made, there might be thinking beings living in our world, but made differently from us, who would think that space has more or less than three dimensions. Has not M. de Cyon said that Japanese mice, having only two pairs of semicircular canals, think that space has two dimensions? Then will not this thinking being, if he is capable of constructing a physical system, make a system of two or four dimensions, which yet, in a sense, will be the same as ours, since it will be the description of the same world in another language?

It quite seems, indeed, that it would be possible to translate our physics into the language of geometry of four dimensions. Attempting such a translation would be giving oneself a great deal of trouble for little profit, and I will content myself with mentioning Hertz's mechanics, in which something of the kind may be seen. Yet it seems that the translation would always be less simple than the text, and that it would never lose the appearance of a translation, for the language of three dimensions seems the best suited to the description of our world, even though that description may be made, in case of necessity, in another idiom.

Besides, it is not by chance that our distribution board has been formed. There is a connexion between the alarm A1 and the parry B1, that is, a property residing in our intelligence. But why is there this connexion? It is because the parry B1 enables us effectively to defend ourselves against the danger A1, and that is a fact exterior to us, a property of the exterior world. Our distribution board, then, is only the translation of an assemblage of exterior facts; if it has three dimensions, it is because it has adapted itself to a world having certain properties, and the most important of these properties is that there exist natural solids which are clearly displaced in accordance with the laws we call laws of motion of unvarying solids. If, then, the language of three dimensions is that which enables us most easily to describe our world, we must not be surprised. This language is founded on our distribution board, and it is in order to enable us to live in this world that this board has been established.

I have said that we could conceive of thinking beings, living in our world, whose distribution board would have four dimensions, who would, consequently, think in hyperspace. It is not certain, however, that such beings, admitting that they were born, would be able to live and defend themselves against the thousand dangers by which they would be assailed.

VI.

A few remarks in conclusion. There is a striking contrast between the roughness of this primitive geometry which is reduced to what I call a distribution board, and the infinite precision of the geometry of geometricians. And yet the latter is the child of the former, but not of it alone; it required to be fertilized by the faculty we have of constructing mathematical concepts, such, for instance, as that of the group. It was necessary to find among these pure concepts the one that was best adapted to this rough space, whose genesis I have tried to explain in the preceding pages, the space which is common to us and the higher animals.

The evidence of certain geometrical postulates is only, as I have said, our unwillingness to give up very old habits. But these postulates are infinitely precise, while the habits have about them something essentially fluid. As soon as we wish to think, we are bound to have infinitely precise postulates, since this is the only means of avoiding contradiction. But among all the possible systems of postulates, there are some that we shall be unwilling to choose, because they do not accord sufficiently with our habits. However fluid and elastic these may be, they have a limit of elasticity.

It will be seen that though geometry is not an experimental science, it is a science born in connexion with experience; that we have created the space it studies, but adapting it to the world in which we live. We have chosen the most convenient space, but experience guided our choice. As the choice was unconscious, it appears to be imposed upon us. Some say that it is imposed by experience, and others that we are born with our space ready-made. After the preceding considerations, it will be seen what proportion of truth and of error there is in these two opinions.

In this progressive education which has resulted

in the construction of space, it is very difficult to determine what is the share of the individual and what of the race. To what extent could one of us, transported from his birth into an entirely different world, where, for instance, there existed bodies displaced in accordance with the laws of motion of non-Euclidian solids—to what extent, I say, would he be able to give up the ancestral space in order to build up an entirely new space?

The share of the race seems to preponderate largely, and yet if it is to it that we owe the rough space, the fluid space of which I spoke just now, the space of the higher animals, is it not to the unconscious experience of the individual that we owe the infinitely precise space of the geometrician? This is a question that is not easy of solution. I would mention, however, a fact which shows that the space bequeathed to us by our ancestors still preserves a certain plasticity. Certain hunters learn to shoot fish under the water, although the image of these fish is raised by refraction; and, moreover, they do it instinctively. Accordingly they have learnt to modify their ancient instinct of direction, or, if you will, to substitute for the association A1, B1, another association A1, B2, because experience has shown them that the former does not succeed.

II.

MATHEMATICAL DEFINITIONS AND EDUCATION.

1. I have to speak here of general definitions in mathematics. At least that is what the title of the chapter says, but it will be impossible for me to confine myself to the subject as strictly as the rule of unity of action demands. I shall not be able to treat it without speaking to some extent of other allied questions, and I must ask your kind forgiveness if I am thus obliged from time to time to walk among the flower-beds to right or left.

What is a good definition? For the philosopher or the scientist, it is a definition which applies to all the objects to be defined, and applies only to them; it is that which satisfies the rules of logic. But in education it is not that; it is one that can be understood by the pupils.

How is it that there are so many minds that are incapable of understanding mathematics? Is there not something paradoxical in this? Here is a science which appeals only to the fundamental principles of logic, to the principle of contradiction, for instance, to what forms, so to speak, the skeleton of our understanding, to what we could not be deprived of without ceasing to think, and yet there are people who find it obscure, and actually they are the majority. That they should be incapable of discovery we can understand, but that they should fail to understand the demonstrations expounded to them, that they should remain blind when they are shown a light that seems to us to shine with a pure brilliance, it is this that is altogether miraculous.

And yet one need have no great experience of examinations to know that these blind people are by no means exceptional beings. We have here a problem that is not easy of solution, but yet must engage the attention of all who wish to devote themselves to education.

What is understanding? Has the word the same meaning for everybody? Does understanding the demonstration of a theorem consist in examining each of the syllogisms of which it is composed in succession, and being convinced that it is correct and conforms to the rules of the game? In the same way, does understanding a definition consist simply in recognizing that the meaning of all the terms employed is already known, and being convinced that it involves no contradiction?

Yes, for some it is; when they have arrived at the conviction, they will say, I understand. But not for the majority. Almost all are more exacting; they want to know not only whether all the syllogisms of a demonstration are correct, but why they are linked together in one order rather than in another. As long as they appear to them engendered by caprice, and not by an intelligence constantly conscious of the end to be attained, they do not think they have understood.

No doubt they are not themselves fully aware of what they require and could not formulate their desire, but if they do not obtain satisfaction, they feel vaguely that something is wanting. Then what happens? At first they still perceive the evidences that are placed before their eyes, but, as they are connected by too attenuated a thread with those that precede and those that follow, they pass without leaving a trace in their brains, and are immediately forgotten; illuminated for a moment, they relapse at once into an eternal night. As they advance further, they will no longer see even this ephemeral light, because the theorems depend one upon another, and those they require have been forgotten. Thus it is that they become incapable of understanding mathematics.

It is not always the fault of their instructor. Often their intellect, which requires to perceive the connecting thread, is too sluggish to seek it and find it. But in order to come to their assistance, we must first of all thoroughly understand what it is that stops them.

Others will always ask themselves what use it is. They will not have understood, unless they find around them, in practice or in nature, the object of such and such a mathematical notion. Under each word they wish to put a sensible image; the definition must call up this image, and at each stage of the demonstration they must see it being transformed and evolved. On this condition only will they understand and retain what they have understood. These often deceive themselves: they do not listen to the reasoning, they look at the figures; they imagine that they have understood when they have only seen.

2. What different tendencies we have here! Are we to oppose them, or are we to make use of them? And if we wish to oppose them, which are we to favour? Are we to show those who content themselves with the pure logic that they have only seen one side of the matter, or must we tell those who are not so easily satisfied that what they demand is not necessary?

In other words, should we constrain young people to change the nature of their minds? Such an attempt would be useless; we do not possess the philosopher's stone that would enable us to transmute the metals entrusted to us one into the other. All that we can do is to work them, accommodating ourselves to their properties.

Many children are incapable of becoming mathematicians who must none the less be taught mathematics; and mathematicians themselves are not all cast in the same mould. We have only to read their works to distinguish among them two kinds of minds—logicians like Weierstrass, for instance, and intuitionists like Riemann. There is the same difference among our students. Some prefer to treat their problems "by analysis," as they say, others "by geometry."

It is quite useless to seek to change anything in this, and besides, it would not be desirable. It is well that there should be logicians and that there should be intuitionists. Who would venture to say whether he would prefer that Weierstrass had never written or that there had never been a Riemann? And so we must resign ourselves to the diversity of minds, or rather we must be glad of it.

3. Since the word understand has several meanings, the definitions that will be best understood by some are not those that will be best suited to others. We have those who seek to create an image, and those who restrict themselves to combining empty forms, perfectly intelligible, but purely intelligible, and deprived by abstraction of all matter.

I do not know whether it is necessary to quote any examples, but I will quote some nevertheless, and, first, the definition of fractions will furnish us with an extreme example. In the primary schools, when they want to define a fraction, they cut up an apple or a pie. Of course this is done only in imagination and not in reality, for I do not suppose the budget of primary education would allow such an extravagance. In the higher normal school, on the contrary, or in the universities, they say: a fraction is the combination of two whole numbers separated by a horizontal line. By conventions they define the operations that these symbols can undergo; they demonstrate that the rules of these operations are the same as in the calculation of whole numbers; and, lastly, they establish that multiplication of the fraction by the denominator, in accordance with these rules, gives the numerator. This is very well, because it is addressed to young people long since familiarized with the notion of fractions by dint of cutting up apples and other objects, so that their mind, refined by a considerable mathematical education, has, little by little, come to desire a purely logical definition. But what would be the consternation of the beginner to whom we attempted to offer it?

Such, also, are the definitions to be found in a

book that has been justly admired and has received several awards of merit—Hilbert's "Grundlagen der Geometrie." Let us see how he begins. "Imagine three systems of THINGS, which we will call points, straight lines, and planes." What these "things" are we do not know, and we do not need to know—it would even be unfortunate that we should seek to know; all that we have the right to know about them is that we should learn their axioms, this one, for instance: "Two different points always determine a straight line," which is followed by this commentary: "Instead of determine we may say that the straight line passes through these two points, or that it joins these two points, or that the two points are situated on the straight line." Thus "being situated on a straight line" is simply defined as synonymous with "determining a straight line." Here is a book of which I think very highly, but which I should not recommend to a schoolboy. For the matter of that I might do it without fear; he would not carry his reading very far.

I have taken extreme examples, and no instructor would dream of going so far. But, even though he comes nowhere near such models, is he not still exposed to the same danger?

We are in a class of the fourth grade. The teacher is dictating: "A circle is the position of the points in a plane which are the same distance from an interior point called the centre." The good pupil writes this phrase in his copy-book and the bad pupil draws faces, but neither of them understands. Then the teacher takes the chalk and draws a circle on the board. "Ah," think the pupils, "why didn't he say at once, a circle is a round, and we should have understood." No doubt it is the teacher who is right. The pupils' definition would have been of no value, because it could not have been used for any demonstration, and chiefly because it could not have given them the salutary habit of analyzing their conceptions. But they should be made to see that they do not understand what they think they understand, and brought to realize the roughness of their primitive concept, and to be anxious themselves that it should be purified and refined.

4. I shall return to these examples; I only wished to show the two opposite conceptions. There is a violent contrast between them, and this contrast is explained by the history of the science. If we read a book written fifty years ago, the greater part of the arguments appear to us devoid of exactness.

At that period they assumed that a continuous function cannot change its sign without passing through zero, but to-day we prove it. They assumed that the ordinary rules of calculus are applicable to incommensurable numbers; to-day we prove it. They assumed many other things that were sometimes untrue.

They trusted to intuition, but intuition cannot give us exactness, nor even certainty, and this has been recognized more and more. It teaches us, for instance, that every curve has a tangent—that is to say, that every continuous function has a derivative—and that is untrue. As certainty was required, it has been necessary to give less and less place to intuition.

How has this necessary evolution come about? It was not long before it was recognized that exactness cannot be established in the arguments unless it is first introduced into the definitions.

For a long time the objects that occupied the attention of mathematicians were badly defined. They thought they knew them because they represented them by their senses or their imagination, but they had only a rough image, and not a precise idea such as reasoning can take hold of.

It is to this that the logicians have had to apply their efforts, and similarly for incommensurable numbers.

The vague idea of continuity which we owe to intuition has resolved itself into a complicated system of inequalities bearing on whole numbers. Thus it is that all those difficulties which terrified our ancestors when they reflected upon the foundations of the infinitesimal calculus have finally vanished.

In analysis to-day there is no longer anything but whole numbers, or finite or infinite systems of whole numbers, bound together by a network of equalities and inequalities. Mathematics, as it has been said, has been arithmetized.

5. But we must not imagine that the science of mathematics has attained to absolute exactness without making any sacrifice. What it has gained in exactness it has lost in objectivity. It is by withdrawing from reality that it has acquired this perfect purity. We can now move freely over its whole domain, which formerly bristled with obstacles. But these obstacles have not disappeared; they have only been removed to the frontier, and will have to be conquered again if we wish to cross the frontier and penetrate into the realms of practice.

We used to possess a vague notion, formed of incongruous elements, some *a priori* and others derived from more or less digested experiences, and we imagined we knew its principal properties by intuition. To-day we reject the empirical element and preserve only the *a priori* ones. One of the properties serves as definition, and all the others are deduced from it by exact reasoning. This is very well, but it still remains to prove that this property, which has become a definition, belongs to the real objects taught us by experience, from which we had drawn our vague intuitive notion. In order to prove it we shall certainly have to appeal to experience or make an effort of intuition; and if we cannot prove it, our theorems will be perfectly exact but perfectly useless.

Logic sometimes breeds monsters. For half a century there has been springing up a host of weird functions, which seem to strive to have as little resemblance as possible to honest functions that are of some use. No more continuity, or else continuity but no derivatives, etc. More than this, from the point of view of logic, it is these strange functions that are the most general; those that are met without being looked for no longer appear as more than a particular case, and they have only quite a little corner left them.

Formerly, when a new function was invented, it was in view of some practical end. To-day they are invented on purpose to show our ancestors' reasonings at fault, and we shall never get anything more than that out of them.

If logic were the teacher's only guide, he would have to begin with the most general, that is to say, with the most weird, functions. He would have to set the beginner to wrestle with this collection of monstrosities. If you don't do so, the logicians might say, you will only reach exactness by stages.

6. Possibly this may be true, but we cannot take such poor account of reality, and I do not mean merely the reality of the sensible world, which has its value nevertheless, since it is for battling with it that nine-tenths of our pupils are asking for arms. There is a more subtle reality which constitutes the life of mathematical entities, and is something more than logic.

Our body is composed of cells, and the cells of atoms, but are these cells and atoms the whole reality of the human body? Is not the manner in which these cells are adjusted, from which results the unity of the individual, also a reality, and of much greater interest?

Would a naturalist imagine that he had an adequate knowledge of the elephant if he had never studied the animal except through a microscope?

It is the same in mathematics. When the logician has resolved each demonstration into a host of elementary operations, all of them correct, he will not yet be in possession of the whole reality; that indefinable something that constitutes the unity of the demonstration will still escape him completely.

What good is it to admire the mason's work in the edifices erected by great architects, if we cannot understand the general plan of the master? Now pure logic cannot give us this view of the whole; it is to intuition we must look for it.

Take, for instance, the idea of the continuous function. To begin with, it is only a perceptible image, a line drawn with chalk on a blackboard. Little by little it is purified; it is used for constructing a complicated system of inequalities which reproduces all the lines of the original image; when the work is quite finished, the *centering* is removed, as it is after the construction of an arch; this crude representation is henceforth a useless support, and disappears, and there remains only the edifice itself, irreproachable in the eyes of the logician. And yet, if the instructor did not recall the original image, if he did not replace the *centering* for a moment, how would the pupil guess by what caprice all these inequalities had been scaffolded in this way one upon another? The definition would be logically correct, but it would not show him the true reality.

7. And so we are obliged to make a step backwards. No doubt it is hard for a master to teach what does not satisfy him entirely, but the satisfaction of the master is not the sole object of education. We have first to concern ourselves with the pupil's state of mind, and what we want it to become.

Zoologists declare that the embryonic development of an animal repeats in a very short period of time the whole history of its ancestors of the geological ages. It seems to be the same with the development of minds. The educator must make the child pass through all that his fathers have passed through, more rapidly, but without missing a stage. On this account, the history of any science must be our first guide.

Our fathers imagined they knew what a fraction was, or continuity, or the area of a curved surface; it is we who have realized that they did not. In the same way our pupils imagine that they know it when they begin to study mathematics seriously. If, without any other preparation, I come and say to them: "No, you do not know it; you do not understand what you imagine you understand; I must demonstrate to you what appears to you evident;" and if, in the demonstration, I rely on premises that seem to them less evident than the conclusion, what will the wretched pupils think? They will think that the science of mathematics is nothing but an arbitrary aggregation of useless subtleties; or they will lose their taste for it; or else they will look upon it as an amusing game, and arrive at a state of mind analogous to that of the Greek sophists.

Later on, on the contrary, when the pupil's mind has been familiarized with mathematical reasoning and ripened by this long intimacy, doubts will spring up of their own accord, and then your demonstration will be welcome. It will arouse new doubts, and questions will present themselves successively to the child, as they presented themselves successively to our fathers, until they reach a point when only perfect exactness will satisfy them. It is not enough to feel doubts about everything; we must know why we doubt.

8. The principal aim of mathematical education is to develop certain faculties of the mind, and among these intuition is not the least precious. It is through it that the mathematical world remains in touch with the real world, and even if pure mathematics could do without it, we should still have to have recourse to it to fill up the gulf that separates the symbol

from reality. The practitioner will always need it, and for every pure geometrician there must be a hundred practitioners.

The engineer must receive a complete mathematical training, but of what use is it to be to him, except to enable him to see the different aspects of things and to see them quickly? He has no time to split hairs. In the complex physical objects that present themselves to him, he must promptly recognize the point where he can apply the mathematical instruments we have put in his hands. How could he do this if we left between the former and the latter that deep gulf dug by the logicians?

9. Beside the future engineers are other less numerous pupils, destined in their turn to become teachers, and so they must go to the very root of the matter; a profound and exact knowledge of first principles is above all indispensable for them. But that is no reason for not cultivating their intuition, for they would form a wrong idea of the science if they never looked at it on more than one side, and, besides, they could not develop in their pupils a quality they did not possess themselves.

For the pure geometrician himself this faculty is necessary: it is by logic that we prove, but by intuition that we discover. To know how to criticize is good, but to know how to create is better. You know how to recognize whether a combination is correct, but much use this will be if you do not possess the art of selecting among all the possible combinations. Logic teaches us that on such and such a road we are sure of not meeting an obstacle;

it does not tell us which is the road that leads to the desired end. For this it is necessary to see the end from afar, and the faculty which teaches us to see is intuition. Without it, the geometrician would be like a writer well up in grammar but destitute of ideas. Now how is this faculty to develop, if, as soon as it shows itself, it is hounded out and proscribed, if we learn to distrust it before we know what good can be got from it?

And here let me insert a parenthesis to insist on the importance of written exercises. Compositions in writing are perhaps not given sufficient prominence in certain examinations. In the *École Polytechnique*, for instance, I am told that insistence on such compositions would close the door to very good pupils who know their subject and understand it very well, and yet are incapable of applying it in the smallest degree. I said just above that the word understand has several meanings. Such pupils only understand in the first sense of the word, and we have just seen that this is not sufficient to make either an engineer or a geometrician. Well, since we have to make a choice, I prefer to choose those who understand thoroughly.

10. But is not the art of exact reasoning also a precious quality that the teacher of mathematics should cultivate above all else? I am in no danger of forgetting it: we must give it attention, and that from the beginning. I should be distressed to see geometry degenerate into some sort of low-grade tachymetrics, and I do not by any means subscribe to the extreme doctrines of certain German professors. But we have sufficient opportunity of training pupils

in correct reasoning in those parts of mathematics in which the disadvantages I have mentioned do not occur. We have long series of theorems in which absolute logic has ruled from the very start and, so to speak, naturally, in which the first geometricians have given us models that we must continually imitate and admire.

It is in expounding the first principles that we must avoid too much subtlety, for there it would be too disheartening, and useless besides. We cannot prove everything, we cannot define everything, and it will always be necessary to draw upon intuition. What does it matter whether we do this a little sooner or a little later, and even whether we ask for a little more or a little less, provided that, making a correct use of the premises it gives us, we learn to reason accurately?

11. Is it possible to satisfy so many opposite conditions? Is it possible especially when it is a question of giving a definition? How are we to find a statement that will at the same time satisfy the inexorable laws of logic and our desire to understand the new notion's place in the general scheme of the science, our need of thinking in images? More often than not we shall not find it, and that is why the statement of a definition is not enough; it must be prepared and it must be justified.

What do I mean by this? You know that it has often been said that every definition implies an axiom, since it asserts the existence of the object defined. The definition, then, will not be justified, from the purely logical point of view, until we have *proved* that it involves no contradiction either in its terms or with the truths previously admitted.

But that is not enough. A definition is stated as a convention, but the majority of minds will revolt if you try to impose it upon them as an *arbitrary* convention. They will have no rest until you have answered a great number of questions.

Mathematical definitions are most frequently, as M. Liard has shown, actual constructions built up throughout of simpler notions. But why should these elements have been assembled in this manner, when a thousand other assemblages were possible? Is it simply caprice? If not, why had this combination more right to existence than any of the others? What need does it fill? How was it foreseen that it would play an important part in the development of the science, that it would shorten our reasoning and our calculations? Is there any familiar object in nature that is, so to speak, its indistinct and rough image?

That is not all. If you give a satisfactory answer to all these questions, we shall realize that the new-comer had the right to be baptized. But the choice of a name is not arbitrary either; we must explain what analogies have guided us, and that if we have given analogous names to different things, these things at least differ only in matter, and have some resemblance in form, that their properties are analogous and, so to speak, parallel.

It is on these terms that we shall satisfy all propensities. If the statement is sufficiently exact to please the logician, the justification will satisfy the intuitionist. But we can do better still. Whenever it is possible, the justification will precede the statement and prepare it. The general statement will be led up to by the study of some particular examples.

One word more. The aim of each part of the statement of a definition is to distinguish the object to be defined from a class of other neighbouring objects. The definition will not be understood until you have shown not only the object defined, but the neighbouring objects from which it has to be distinguished, until you have made it possible to grasp the difference, and have added explicitly your reason for saying this or that in stating the definition.

But it is time to leave generalities and to enquire how the somewhat abstract principles I have been expounding can be applied in arithmetic, in geometry, in analysis, and in mechanics.

ARITHMETIC.

12. We do not have to define the whole number. On the other hand, operations on whole numbers are generally defined, and I think the pupils learn these definitions by heart and attach no meaning to them. For this there are two reasons: first, they are taught them too early, while their mind still feels no need of them; and then these definitions are not satisfactory from the logical point of view. For addition, we cannot find a good one, simply because we must stop somewhere, and cannot define everything. The definition of addition is to say that it consists in adding. All that we can do is to start with a certain number of concrete examples and say, the operation that has just been performed is called addition.

For subtraction it is another matter. It can be defined logically as the inverse operation of addition.

But is that how we should begin? Here, again, we should start with examples, and show by these examples the relation of the two operations. Thus the definition will be prepared and justified.

In the same way for multiplication. We shall take a particular problem; we shall show that it can be solved by adding several equal numbers together; we shall then point out that we arrive at the result quicker by multiplication, the operation the pupils perform already by rote, and the logical definition will spring from this quite naturally.

We shall define division as the inverse operation of multiplication; but we shall begin with an example drawn from the familiar notion of sharing, and we shall show by this example that multiplication reproduces the dividend.

There remain the operations on fractions. There is no difficulty except in the case of multiplication. The best way is first to expound the theory of proportions, as it is from it alone that the logical definition can spring. But, in order to gain acceptance for the definitions that are met with at the start in this theory, we must prepare them by numerous examples drawn from classical problems of the rule of three, and we shall be careful to introduce fractional data. We shall not hesitate, either, to familiarize the pupils with the notion of proportion by geometrical figures; either appealing to their recollection if they have already done any geometry, or having recourse to direct intuition if they have not, which, moreover, will prepare them to do it. I would add, in conclusion, that after having defined the multiplication of fractions, we must justify this definition by demonstration that it is commutative, associative, and distributive, making it quite clear to the listeners that the verification has been made in order to justify the definition.

We see what part is played in all this by geometrical figures, and this part is justified by the philosophy and the history of the science. If arithmetic had remained free from all intermixture with geometry, it would never have known anything but the whole number. It was in order to adapt itself to the requirements of geometry that it discovered something else.

Geometry.

In geometry we meet at once the notion of the straight line. Is it possible to define the straight line? The common definition, the shortest path from one point to another, does not satisfy me at all. I should start simply with the *ruler*, and I should first show the pupil how we can verify a ruler by revolving it. This verification is the true definition of a straight line, for a straight line is an axis of rotation. We should then show him how to verify the ruler by sliding it, and we should have one of the most important properties of a straight line. As for that other property, that of being the shortest path from one point to another, it is a theorem that can be demonstrated apodeictically, but the demonstration is too advanced to find a place in secondary education. It will be better to show that a ruler previously verified can be applied to a taut thread. We must not hesitate, in the presence of difficulties of this kind, to multiply the axioms, justifying them by rough examples.

Some axioms we must admit; and if we admit a

few more than is strictly necessary, the harm is not great. The essential thing is to learn to reason exactly with the axioms once admitted. Uncle Sarcey, who loved to repeat himself, often said that the audience at a theatre willingly accepts all the postulates imposed at the start, but that once the curtain has gone up it becomes inexorable on the score of logic. Well, it is just the same in mathematics.

For the circle we can start with the compass. The pupils will readily recognize the curve drawn. We shall then point out to them that the distance of the two points of the instrument remains constant, that one of these points is fixed and the other movable, and we shall thus be led naturally to the logical definition.

The definition of a plane implies an axiom, and we must not attempt to conceal the fact. Take a drawing-board and point out how a movable ruler can be applied constantly to the board, and that while still retaining three degrees of freedom. We should compare this with the cylinder and the cone, surfaces to which a straight line cannot be applied unless we allow it only two degrees of freedom. Then we should take three drawing-boards, and we should show first that they can slide while still remaining in contact with one another, and that with three degrees of freedom. And lastly, in order to distinguish the plane from the sphere, that two of these boards that can be applied to a third can also be applied to one another.

Perhaps you will be surprised at this constant use of movable instruments. It is not a rough artifice, and it is much more philosophical than it would appear at first sight. What is geometry for the philosopher? It is the study of a group. And what group? That of the movements of solid bodies. How are we to define this group, then, without making some solid bodies move?

Are we to preserve the classical definition of parallels, and say that we give this name to two straight lines, situated in the same plane, which, being produced ever so far, never meet? No, because this definition is negative, because it cannot be verified by experience, and cannot consequently be regarded as an immediate datum of intuition, but chiefly because it is totally foreign to the notion of group and to the consideration of the motion of solid bodies, which is, as I have said, the true source of geometry. Would it not be better to define first the rectilineal transposition of an invariable figure as a motion in which all the points of this figure have rectilineal trajectories, and to show that such a transposition is possible, making a square slide on a ruler? From this experimental verification, raised to the form of an axiom, it would be easy to educe the notion of parallel and Euclid's postulate itself.

MECHANICS.

I need not go back to the definition of velocity or of acceleration or of the other kinematic notions: they will be more properly connected with ideas of space and time, which alone they involve.

On the contrary, I will dwell on the dynamic notions of force and mass.

There is one thing that strikes me, and that is, how

far young people who have received a secondary education are from applying the mechanical laws they have been taught to the real world. It is not only that they are incapable of doing so, but they do not even think of it. For them the world of science and that of reality are shut off in water-tight compartments. It is not uncommon to see a well-dressed man, probably a university man, sitting in a carriage and imagining that he is helping it on by pushing on the dash-board, and that in disregard of the principle of action and reaction.

If we try to analyze the state of mind of our pupils, this will surprise us less. What is for them the true definition of force? Not the one they repeat, but the one that is hidden away in a corner of their intellect, and from thence directs it all. This is their definition: Forces are arrows that parallelograms are made of; these arrows are imaginary things that have nothing to do with anything that exists in nature. This would not happen if they were shown forces in reality before having them represented by arrows.

How are we to define force? If we want a logical definition, there is no good one, as I think I have shown satisfactorily elsewhere. There is the anthropomorphic definition, the sensation of muscular effort; but this is really too crude, and we cannot extract anything useful from it.

This is the course we ought to pursue. First, in order to impart a knowledge of the genus force, we must show, one after the other, all the species of this genus. They are very numerous and of great variety. There is the pressure of liquids on the sides of the vessels in which they are contained, the tension of cords, the elasticity of a spring, gravity that acts on all the molecules of a body, friction, the normal mutuai action and reaction of two solids in contact.

This is only a qualitative definition; we have to learn to measure a force. For this purpose we shall show first that we can replace one force by another without disturbing the equilibrium, and we shall find the first example of this substitution in the balance and Borda's double scales. Then we shall show that we can replace a weight not only by another weight, but by forces of different nature; for example, Prony's dynamometer break enables us to replace a weight by friction.

From all this arises the notion of the equivalence of two forces.

We must also define the direction of a force. If a force F is equivalent to another force F^1 that is applied to the body we are dealing with through the medium of a taut cord, in such a way that F can be replaced by F^1 without disturbing the equilibrium, then the point of attachment of the cord will be, by definition, the point of application of the force F^1 and that of the equivalent force F, and the direction of the cord will be the direction of the force F^1 and also that of the equivalent force F.

From this we shall pass to the comparison of the magnitude of forces. If one force can replace two others of the same direction, it must be equal to their sum, and we shall show, for instance, that a weight of 20 ounces can replace two weights of 10 ounces.

But this is not all. We know now how to compare the intensity of two forces which have the same direction and the same point of application, but we have to learn to do this when the directions are different. For this purpose we imagine a cord stretched by a weight and passing over a pulley; we say that the tension of the two portions of the cord is the same, and equal to the weight.

Here is our definition. It enables us to compare the tensions of our two portions, and, by using the preceding definitions, to compare two forces of any kind having the same direction as these two portions. We have to justify it by showing that the tension of the last portion remains the same for the same weight, whatever be the number and the disposition of the pulleys. We must then complete it by showing that this is not true unless the pulleys are without friction.

Once we have mastered these definitions we must show that the point of application, the direction, and the intensity are sufficient to determine a force; that two forces for which these three elements are the same are *always* equivalent, and can *always* be replaced one by the other, either in equilibrium or in motion, and that whatever be the other forces coming into play.

We must show that two concurrent forces can always be replaced by a single resultant force, and that *this resultant remains the same* whether the body is in repose or in motion, and whatever be the other forces applied to it.

Lastly, we must show that forces defined as we have defined them satisfy the principle of the equality of action and reaction.

All this we learn by experiment, and by experiment alone.

It will be sufficient to quote some common experiments that the pupils make every day without being aware of it, and to perform before them a small number of simple and well-selected experiments.

It is not until we have passed through all these roundabout ways that we can represent forces by arrows, and even then I think it would be well, from time to time, as the argument develops, to come back from the symbol to the reality. It would not be difficult, for instance, to illustrate the parallelogram of forces with the help of an apparatus composed of three cords passing over pulleys, stretched by weights, and producing equilibrium by pulling on the same point.

Once we know force, it is easy to define mass. This time the definition must be borrowed from dynamics. We cannot do otherwise, since the end in view is to make clear the distinction between mass and weight. Here, again, the definition must be prepared by experiments. There is, indeed, a machine that seems to be made on purpose to show what mass is, and that is Atwood's machine. Besides this we shall recall the laws of falling bodies, and how acceleration of gravity is the same for heavy as for light bodies, and varies according to latitude, etc.

Now if you tell me that all the methods I advocate have long since been applied in schools, I shall be more pleased than surprised to hear it. I know that on the whole our mathematical education is good; I do not wish to upset it, and should even be distressed at this result; I only desire gradual, progressive improvements. This education must not undergo sudden variations at the capricious breath of ephemeral fashions. In such storms its high educative value would soon founder. A good and sound logic must continue to form its foundation. Definition by example is always necessary, but it must prepare the logical definition and not take its place; it must at least make its want felt in cases where the true logical definition cannot be given to any purpose except in higher education.

You will understand that what I have said here in no sense implies the abandonment of what I have written elsewhere. I have often had occasion to criticize certain definitions which I advocate to-day. These criticisms hold good in their entirety; the definitions can only be provisional, but it is through them that we must advance.

III.

MATHEMATICS AND LOGIC.

INTRODUCTION.

CAN mathematics be reduced to logic without having to appeal to principles peculiar to itself? There is a whole school full of ardour and faith who make it their business to establish the possibility. They have their own special language, in which words are used no longer, but only signs. This language can be understood only by the few initiated, so that the vulgar are inclined to bow before the decisive affirmations of the adepts. It will, perhaps, be useful to examine these affirmations somewhat more closely, in order to see whether they justify the peremptory tone in which they are made.

But in order that the nature of the question should be properly understood, it is necessary to enter into some historical details, and more particularly to review the character of Cantor's work.

The notion of infinity had long since been introduced into mathematics, but this infinity was what philosophers call a *becoming*. Mathematical infinity was only a quantity susceptible of growing beyond all limit; it was a variable quantity of which it could not be said that it *had passed*, but only that it *would pass*, all limits.

Cantor undertook to introduce into mathematics an *actual infinity*—that is to say, a quantity which is not only susceptible of passing all limits, but which is regarded as having already done so. He set himself such questions as these: Are there more points in space than there are whole numbers? Are there more points in space than there are points in a plane? etc.

Then the number of whole numbers, that of points in space, etc., constitutes what he terms a *transfinite cardinal number*—that is to say, a cardinal number greater than all the ordinary cardinal numbers. And he amused himself by comparing these transfinite cardinal numbers, by arranging in suitable order the elements of a whole which contains an infinite number of elements; and he also imagined what he terms transfinite ordinal numbers, on which I will not dwell further.

Many mathematicians have followed in his tracks, and have set themselves a series of questions of the same kind. They have become so familiar with transfinite numbers that they have reached the point of making the theory of finite numbers depend on that of Cantor's cardinal numbers. In their opinion, if we wish to teach arithmetic in a truly logical way, we ought to begin by establishing the general properties of the transfinite cardinal numbers, and then distinguish from among them quite a small class, that of the ordinary whole numbers. Thanks to this roundabout proceeding, we might succeed in proving all the propositions relating to this small class (that is to say, our whole arithmetic and algebra) without making use of a single principle foreign to logic.

This method is evidently contrary to all healthy

psychology. It is certainly not in this manner that the human mind proceeded to construct mathematics, and I imagine, too, its authors do not dream of introducing it into secondary education. But is it at least logical, or, more properly speaking, is it accurate? We may well doubt it.

Nevertheless, the geometricians who have employed it are very numerous. They have accumulated formulas and imagined that they rid themselves of all that is not pure logic by writing treatises in which the formulas are no longer interspersed with explanatory text, as in the ordinary works on mathematics, but in which the text has disappeared entirely.

Unfortunately, they have arrived at contradictory results, at what are called the *Cantorian antinomies*, to which we shall have occasion to return. These contradictions have not discouraged them, and they have attempted to modify their rules, in order to dispose of those that had already appeared, but without gaining any assurance by so doing that no new ones would appear.

It is time that these exaggerations were treated as they deserve. I have no hope of convincing these logicians, for they have lived too long in this atmosphere. Besides, when we have refuted one of their demonstrations, we are quite sure to find it cropping up again with insignificant changes, and some of them have already risen several times from their ashes. Such in old times was the Lernæan hydra, with its famous heads that always grew again. Hercules was successful because his hydra had only nine heads (unless, indeed, it was eleven), but in this case there are too many, they are in England, in Germany, in Italy, and in France, and he would be forced to abandon the task. And so I appeal only to unprejudiced people of common sense.

I

In these latter years a large number of works have been published on pure mathematics and the philosophy of mathematics, with a view to disengaging and isolating the logical elements of mathematical reasoning. These works have been analyzed and expounded very lucidly by M. Couturat in a work entitled "Les Principes des Mathématiques."

In M. Couturat's opinion the new works, and more particularly those of Mr. Russell and Signor Peano, have definitely settled the controversy so long in dispute between Leibnitz and Kant. They have shown that there is no such thing as an *a priori* synthetic judgment (the term employed by Kant to designate the judgments that can neither be demonstrated analytically, nor reduced to identity, nor established experimentally); they have shown that mathematics is entirely reducible to logic, and that intuition plays no part in it whatever.

This is what M. Couturat sets forth in the work I have just quoted. He also stated the same opinions even more explicitly in his speech at Kant's jubilee; so much so that I overheard my neighbour whisper: "It's quite evident that this is the centenary of Kant's *death*."

Can we subscribe to this decisive condemnation? I do not think so, and I will try to show why.

II.

What strikes us first of all in the new mathematics is its purely formal character. "Imagine," says Hilbert, "three kinds of *things*, which we will call points, straight lines, and planes; let us agree that a straight line shall be determined by two points, and that, instead of saying that this straight line is determined by these two points, we may say that it passes through these two points, or that these two points are situated on the straight line." What these *things* are, not only do we not know, but we must not seek to know. It is unnecessary, and any one who had never seen either a point or a straight line or a plane could do geometry just as well as we can. In order that the words *pass through* or the words *be situated on* should not call up any image in our minds, the former is merely regarded as the synonym of *be determined*, and the latter of *determine*.

Thus it will be readily understood that, in order to demonstrate a theorem, it is not necessary or even useful to know what it means. We might replace geometry by the *reasoning piano* imagined by Stanley Jevons; or, if we prefer, we might imagine a machine where we should put in axioms at one end and take out theorems at the other, like that legendary machine in Chicago where pigs go in alive and come out transformed into hams and sausages. It is no more necessary for the mathematician than it is for these machines to know what he is doing.

I do not blame Hilbert for this formal character of his geometry. He was bound to tend in this direction, given the problem he set himself. He wished to reduce to a minimum the number of the fundamental axioms of geometry, and to make a complete enumeration of them. Now, in the arguments in which our mind remains active, in those in which intuition still plays a part, in the living arguments, so to speak, it is difficult not to introduce an axiom or a postulate that passes unnoticed. Accordingly, it was not till he had reduced all geometrical arguments to a purely mechanical form that he could be certain of having succeeded in his design and accomplished his work.

What Hilbert had done for geometry, others have tried to do for arithmetic and analysis. Even if they had been entirely successful, would the Kantians be finally condemned to silence? Perhaps not, for it is certain that we cannot reduce mathematical thought to an empty form without mutilating it. Even admitting that it has been established that all theorems can be deduced by purely analytical processes, by simple logical combinations of a finite number of axioms, and that these axioms are nothing but conventions, the philosopher would still retain the right to seek the origin of these conventions, and to ask why they were judged preferable to the contrary conventions.

And, further, the logical correctness of the arguments that lead from axioms to theorems is not the only thing we have to attend to. Do the rules of perfect logic constitute the whole of mathematics? As well say that the art of the chess-player reduces itself to the rules for the movement of the pieces. A selection must be made out of all the constructions that can be combined with the materials furnished by logic. The true geometrician makes this selection judiciously, because he is guided by a sure instinct, or by some vague consciousness of I know not what profounder and more hidden geometry, which alone gives a value to the constructed edifice.

To seek the origin of this instinct, and to study the laws of this profound geometry which can be felt but not expressed, would be a noble task for the philosophers who will not allow that logic is all. But this is not the point of view I wish to take, and this is not the way I wish to state the question. This instinct I have been speaking of is necessary to the discoverer, but it seems at first as if we could do without it for the study of the science once created. Well, what I want to find out is, whether it is true that once the principles of logic are admitted we can, I will not say discover, but demonstrate all mathematical truths without making a fresh appeal to intuition.

III.

To this question I formerly gave a negative answer. (See "Science et Hypothèse," Chapter I.) Must our answer be modified by recent works? I said no, because "the principle of complete induction" appeared to me at once necessary to the mathematician, and irreducible to logic. We know the statement of the principle: "If a property is true of the number 1, and if it is established that it is true of $n+1$ provided it is true of n, it will be true of all whole numbers." I recognized in this the typical mathematical argument. I did not mean to say, as has been supposed, that all mathematical arguments can be reduced to an application of this principle.

Examining these arguments somewhat closely, we should discover the application of many other similar principles, offering the same essential characteristics. In this category of principles, that of complete induction is only the simplest of all, and it is for that reason that I selected it as a type.

The term principle of complete induction which has been adopted is not justifiable. This method of reasoning is none the less a true mathematical induction itself, which only differs from the ordinary induction by its certainty.

IV.

Definitions and Axioms.

The existence of such principles is a difficulty for the inexorable logicians. How do they attempt to escape it? The principle of complete induction, they say, is not an axiom properly so called, or an *a priori* synthetic judgment; it is simply the definition of the whole number. Accordingly it is a mere convention. In order to discuss this view, it will be necessary to make a close examination of the relations between definitions and axioms.

We will first refer to an article by M. Couturat on mathematical definitions which appeared in *l'Enseignement Mathématique*, a review published by Gauthier-Villars and by Georg in Geneva. We find a distinction between *direct definition* and *definition by postulates.*

"Definition by postulates," says M. Couturat, "applies not to a single notion, but to a system of notions; it consists in enumerating the fundamental relations that unite them, which make it possible to demonstrate all their other properties: these relations are postulates . . ."

If we have previously defined all these notions with one exception, then this last will be by definition the object which verifies these postulates.

Thus certain indemonstrable axioms of mathematics would be nothing but disguised definitions. This point of view is often legitimate, and I have myself admitted it, for instance, in regard to Euclid's postulate.

The other axioms of geometry are not sufficient to define distance completely. Distance, then, will be by definition, the one among all the magnitudes which satisfy the other axioms, that is of such a nature as to make Euclid's postulate true.

Well, the logicians admit for the principle of complete induction what I admit for Euclid's postulate, and they see nothing in it but a disguised definition.

But to give us this right, there are two conditions that must be fulfilled. John Stuart Mill used to say that every definition implies an axiom, that in which we affirm the existence of the object defined. On this score, it would no longer be the axiom that might be a disguised definition, but, on the contrary, the definition that would be a disguised axiom. Mill understood the word existence in a material and empirical sense; he meant that in defining a circle we assert that there are round things in nature.

In this form his opinion is inadmissible. Mathematics is independent of the existence of material objects. In mathematics the word exist can only have one meaning; it signifies exemption from contradiction. Thus rectified, Mill's thought becomes accurate. In defining an object, we assert that the definition involves no contradiction.

If, then, we have a system of postulates, and if we can demonstrate that these postulates involve no contradiction, we shall have the right to consider them as representing the definition of one of the notions found among them. If we cannot demonstrate this, we must admit it without demonstration, and then it will be an axiom. So that if we wished to find the definition behind the postulate, we should discover the axiom behind the definition.

Generally, for the purpose of showing that a definition does not involve any contradiction, we proceed *by example*, and try to form an example of an object satisfying the definition. Take the case of a definition by postulates. We wish to define a notion A, and we say that, by definition, an A is any object for which certain postulates are true. If we can demonstrate directly that all these postulates are true of a certain object B, the definition will be justified, and the object B will be an *example* of A. We shall be certain that the postulates are not contradictory, since there are cases in which they are all true at once.

But such a direct demonstration by example is not always possible. Then, in order to establish that the postulates do not involve contradiction, we must picture all the propositions that can be deduced from these postulates considered as premises, and show that among these propositions there are no two of which one is the contradiction of the other. If the number of these propositions is finite, a direct verification is possible; but this is a case that is not frequent, and, moreover, of little interest.

If the number of the propositions is infinite, we can no longer make this direct verification. We must then have recourse to processes of demonstration, in which we shall generally be forced to invoke that very principle of complete induction that we are attempting to verify.

I have just explained one of the conditions which the logicians were bound to satisfy, *and we shall see further on that they have not done so.*

V.

There is a second condition. When we give a definition, it is for the purpose of using it.

Accordingly, we shall find the word defined in the text that follows. Have we the right to assert, of the object represented by this word, the postulate that served as definition? Evidently we have, if the word has preserved its meaning, if we have not assigned it a different meaning by implication. Now this is what sometimes happens, and it is generally difficult to detect it. We must see how the word was introduced into our text, and whether the door through which it came does not really imply a different definition from the one enunciated.

This difficulty is encountered in all applications of mathematics. The mathematical notion has received a highly purified and exact definition, and for the pure mathematician all hesitation has disappeared. But when we come to apply it, to the physical sciences, for instance, we are no longer dealing with this pure notion, but with a concrete object which is often only a rough image of it. To say that this object satisfies the definition, even approximately, is to enunciate a new truth, which has no longer the character of a conventional postulate, and that experience alone can establish beyond a doubt.

But, without departing from pure mathematics, we still meet with the same difficulty. You give a subtle definition of number, and then, once the definition has been given, you think no more about it, because in reality it is not your definition that has taught you what a number is, you knew it long before, and when you come to write the word number farther on, you give it the same meaning as anybody else. In order to know what this meaning is, and if it is indeed the same in this phrase and in that, we must see how you have been led to speak of number and to introduce the word into the two phrases. I will not explain my point any further for the moment, for we shall have occasion to return to it.

Thus we have a word to which we have explicitly given a definition A. We then proceed to make use of it in our text in a way which implicitly supposes another definition B. It is possible that these two definitions may designate the same object, but that such is the case is a new truth that must either be demonstrated or else admitted as an independent axiom.

We shall see further on that the logicians have not fulfilled this second condition any better than the first.

VI.

The definitions of number are very numerous and of great variety, and I will not attempt to enumerate even their names and their authors. We must not be surprised that there are so many. If any one of them was satisfactory we should not get any new ones. If each new philosopher who has applied himself to the question has thought it necessary to invent another, it is because he was not satisfied with those of his predecessors; and if he was not satisfied, it was because he thought he detected a *petitio principii.*

I have always experienced a profound sentiment of uneasiness in reading the works devoted to this problem. I constantly expect to run against a *petitio principii*, and when I do not detect it at once I am afraid that I have not looked sufficiently carefully.

The fact is that it is impossible to give a definition without enunciating a phrase, and difficult to enunciate a phrase without putting in a name of number, or at least the word several, or at least a word in the plural. Then the slope becomes slippery, and every moment we are in danger of falling into the *petitio principii.*

I will concern myself in what follows with those only of these definitions in which the *petitio principii* is most skilfully concealed.

VII.

Pasigraphy.

The symbolical language created by Signor Peano plays a very large part in these new researches. It is capable of rendering some service, but it appears to me that M. Couturat attaches to it an exaggerated importance that must have astonished Peano himself.

The essential element of this language consists in certain algebraical signs which represent the conjunctions: if, and, or, therefore. That these signs may be convenient is very possible, but that they should be destined to change the face of the whole philosophy is quite another matter. It is difficult to admit that the word *if* acquires, when written ɔ, a virtue it did not possess when written if.

This invention of Peano was first called *pasigraphy*, that is to say, the art of writing a treatise on mathematics without using a single word of the ordinary language. This name defined its scope most exactly. Since then it has been elevated to a more exalted dignity, by having conferred upon it the title of *logistic*. The same word is used, it appears, in the *École de Guerre* to designate the art of the quartermaster, the art of moving and quartering troops.* But no confusion need be feared, and we see at once that the new name implies the design of revolutionizing logic.

We may see the new method at work in a mathematical treatise by Signor Burali-Forti entitled "*Una Questione sui Numeri transfiniti*" (An Enquiry concerning transfinite Numbers), included in Volume XI. of the "*Rendiconti del circolo matematico di Palermo*" (Reports of the mathematical club of Palermo).

I will begin by saying that this treatise is very interesting, and, if I take it here as an example, it

* In the French the confusion is with "*logistique*," the art of the "maréchal des *logis*," or quartermaster. In English the possibility of confusion does not arise.

is precisely because it is the most important of all that have been written in the new language. Besides, the uninitiated can read it, thanks to an interlined Italian translation.

What gives importance to this treatise is the fact that it presented the first example of those antinomies met with in the study of transfinite numbers, which have become, during the last few years, the despair of mathematicians. The object of this note, says Signor Burali-Forti, is to show that there can be two transfinite (ordinal) numbers, *a* and *b*, such that *a* is neither equal to, greater than, nor smaller than, *b*.

The reader may set his mind at rest. In order to understand the considerations that will follow, he does not require to know what a transfinite ordinal number is.

Now Cantor had definitely proved that between two transfinite numbers, as between two finite numbers, there can be no relation other than equality or inequality in one direction or the other. But it is not of the matter of this treatise that I desire to speak here; this would take me much too far from my subject. I only wish to concern myself with the form, and I ask definitely whether this form makes it gain much in the way of exactness, and whether it thereby compensates for the efforts it imposes upon the writer and the reader.

To begin with, we find that Signor Burali-Forti defines the number 1 in the following manner:—

$$1 = \iota\, T' \{Ko \frown (u,h) \in (u \in \text{One}\},$$

a definition eminently fitted to give an idea of the number 1 to people who had never heard it before.

I do not understand Peanian well enough to venture to risk a criticism, but I am very much afraid that this definition contains a *petitio principii*, seeing that I notice the figure 1 in the first half and the word One in the second.

However that may be, Signor Burali-Forti starts with this definition, and, after a short calculation, arrives at the equation

$$(27) \qquad 1 \epsilon \text{No},$$

which teaches us that One is a number.

And since I am on the subject of these definitions of the first numbers, I may mention that M. Couturat has also defined both 0 and 1.

What is zero? It is the number of elements in the class nil. And what is the class nil? It is the class which contains none.

To define zero as nil and nil as none is really an abuse of the wealth of language, and so M. Couturat has introduced an improvement into his definition by writing

$$0 = \iota \Lambda : \phi x = \Lambda . \supset . \Lambda = (x \epsilon \phi x),$$

which means in English: zero is the number of the objects that satisfy a condition that is never fulfilled. But as never means *in no case*, I do not see that any very great progress has been made.

I hasten to add that the definition M. Couturat gives of the number 1 is more satisfactory.

One, he says in substance, is the number of the elements of a class in which any two elements are identical.

It is more satisfactory, as I said, in this sense, that in order to define 1, he does not use the word one; on the other hand, he does use the word two.

But I am afraid that if we asked M. Couturat what two is, he would be obliged to use the word one.

VIII.

But let us return to the treatise of Signor Burali-Forti. I said that his conclusions are in direct opposition to those of Cantor. Well, one day I received a visit from M. Hadamard, and the conversation turned upon this antinomy.

"Does not Burali-Forti's reasoning," I said, "seem to you irreproachable?"

"No," he answered; "and, on the contrary, I have no fault to find with Cantor's. Besides, Burali-Forti had no right to speak of the whole of *all* the ordinal numbers."

"Excuse me, he had that right, since he could always make the supposition that

$$\Omega = T'(No, \bar{\epsilon} >).$$

I should like to know who could prevent him. And can we say that an object does not exist when we have called it Ω?"

It was quite useless; I could not convince him (besides, it would have been unfortunate if I had, since he was right). Was it only because I did not speak Peanian with sufficient eloquence? Possibly, but, between ourselves, I do not think so.

Thus, in spite of all this pasigraphical apparatus, the question is not solved. What does this prove? So long as it is merely a question of demonstrating that one is a number, pasigraphy is equal to the task; but if a difficulty presents itself, if there is an antinomy to be resolved, pasigraphy becomes powerless.

IV.

THE NEW LOGICS.

I.

Russell's Logic.

In order to justify its pretensions, logic has had to transform itself. We have seen new logics spring up, and the most interesting of these is Mr. Bertrand Russell's. It seems as if there could be nothing new written about formal logic, and as if Aristotle had gone to the very bottom of the subject. But the field that Mr. Russell assigns to logic is infinitely more extensive than that of the classical logic, and he has succeeded in expressing views on this subject that are original and sometimes true.

To begin with, while Aristotle's logic was, above all, the logic of classes, and took as its starting-point the relation of subject and predicate, Mr. Russell subordinates the logic of classes to that of propositions. The classical syllogism, "Socrates is a man," etc., gives place to the hypothetical syllogism, "If A is true, B is true; now if B is true, C is true, etc." This is, in my opinion, one of the happiest of ideas, for the classical syllogism is easily reduced to the hypothetical syllogism, while the inverse transformation cannot be made without considerable difficulty.

But this is not all. Mr. Russell's logic of propositions is the study of the laws in accordance with which combinations are formed with the conjunctions *if*, *and*, *or*, and the negative *not*. This is a considerable extension of the ancient logic. The properties of the classical syllogism can be extended without any difficulty to the hypothetical syllogism, and in the forms of this latter we can easily recognize the scholastic forms; we recover what is essential in the classical logic. But the theory of the syllogism is still only the syntax of the conjunction *if* and, perhaps, of the negative.

By adding two other conjunctions, *and* and *or*, Mr. Russell opens up a new domain to logic. The signs *and* and *or* follow the same laws as the two signs $\times$ and $+$, that is to say, the commutative, associative, and distributive laws. Thus *and* represents logical multiplication, while *or* represents logical addition. This, again, is most interesting.

Mr. Russell arrives at the conclusion that a false proposition of any kind involves all the other propositions, whether true or false. M. Couturat says that this conclusion will appear paradoxical at first sight. However, one has only to correct a bad mathematical paper to recognize how true Mr. Russell's view is. The candidate often takes an immense amount of trouble to find the first false equation; but as soon as he has obtained it, it is no more than child's play for him to accumulate the most surprising results, some of which may actually be correct.

II.

We see how much richer this new logic is than the classical logic. The symbols have been multiplied and admit of varied combinations, *which are no longer of limited number.* Have we any right to give this extension of meaning to the word *logic*? It would be idle to examine this question, and to quarrel with Mr. Russell merely on the score of words. We will grant him what he asks; but we must not be surprised if we find that certain truths which had been declared to be irreducible to logic, in the old sense of the word, have become reducible to logic, in its new sense, which is quite different.

We have introduced a large number of new notions, and they are not mere combinations of the old. Moreover, Mr. Russell is not deceived on this point, and not only at the beginning of his first chapter—that is to say, his logic of propositions—but at the beginning of his second and third chapters also—that is to say, his logic of classes and relations—he introduces new words which he declares to be undefinable.

And that is not all. He similarly introduces principles which he declares to be undemonstrable. But these undemonstrable principles are appeals to intuition, *a priori* synthetic judgments. We regarded them as intuitive when we met them more or less explicitly enunciated in treatises on mathematics. Have they altered in character because the meaning of the word logic has been extended, and we find them now in a book entitled "Treatise on Logic"? *They have not changed in nature, but only in position.*

III.

Could these principles be considered as disguised definitions? That they should be so, we should require to be able to demonstrate that they involve no contradiction. We should have to establish that, however far we pursue the series of deductions, we shall never be in danger of contradicting ourselves.

We might attempt to argue as follows. We can verify the fact that the operations of the new logic, applied to premises free from contradiction, can only give consequences equally free from contradiction. If then, after n operations, we have not met with contradiction, we shall not meet it any more after $n+1$. Accordingly, it is impossible that there can be a moment when contradiction will *begin*, which shows that we shall never meet it. Have we the right to argue in this way? No, for it would be making complete induction, and we must not forget that *we do not yet know the principle of complete induction.*

Therefore we have no right to regard these axioms as disguised definitions, and we have only one course left. Each one of them, we admit, is a new act of intuition. This is, moreover, as I believe, the thought of Mr. Russell and M. Couturat.

Thus each of the nine undefinable notions and twenty undemonstrable propositions (I feel sure that, if I had made the count, I should have found one or two more) which form the groundwork of the new logic—of the logic in the broad sense—presupposes a new and independent act of our intuition, and why should we not term it a true *a priori* synthetic judgment? On this point everybody seems to be agreed; but what Mr. Russell claims, *and what appears to me doubtful, is that after these appeals to intuition we shall have finished: we shall have no more to make, and we shall be able to construct the whole of mathematics without bringing in a single new element.*

IV.

M. Couturat is fond of repeating that this new logic is quite independent of the idea of number. I will not amuse myself by counting how many instances his statement contains of adjectives of number, cardinal as well as ordinal, or of indefinite adjectives such as several. However, I will quote a few examples:—

"The logical product of *two* or of *several* propositions is......"

"All propositions are susceptible of *two* values only, truth or falsehood."

"The relative product of *two* relations is a relation."

"A relation is established between *two* terms."

Sometimes this difficulty would not be impossible to avoid, but sometimes it is essential. A relation is incomprehensible without two terms. It is impossible to have the intuition of a relation, without having at the same time the intuition of its two terms, and without remarking that they are two, since, for a relation to be conceivable, they must be two and two only.

V.

Arithmetic.

I come now to what M. Couturat calls the *ordinal theory*, which is the groundwork of arithmetic properly so called. M. Couturat begins by enunciating Peano's five axioms, which are independent, as Signor Peano and Signor Padoa have demonstrated.

1. Zero is a whole number.

2. Zero is not the sequent of any whole number.

3. The sequent of a whole number is a whole number. To which it would be well to add: every whole number has a sequent.

4. Two whole numbers are equal if their sequents are equal.

The 5th axiom is the principle of complete induction.

M. Couturat considers these axioms as disguised definitions; they constitute the definition by postulates of zero, of the "sequent," and of the whole number.

But we have seen that, in order to allow of a definition by postulates being accepted, we must be able to establish that it implies no contradiction.

Is this the case here? Not in the very least.

The demonstration cannot be made *by example.* We cannot select a portion of whole numbers—for instance, the three first—and demonstrate that they satisfy the definition.

If I take the series 0, 1, 2, I can readily see that it satisfies axioms 1, 2, 4, and 5; but in order that it should satisfy axiom 3, it is further necessary that 3 should be a whole number, and consequently that the series 0, 1, 2, 3 should satisfy the axioms. We could verify that it satisfies axioms 1, 2, 4, and 5, but axiom 3 requires besides that 4 should be a whole number, and that the series 0, 1, 2, 3, 4 should satisfy the axioms, and so on indefinitely.

It is, therefore, impossible to demonstrate the axioms for some whole numbers without demonstrating them for all, and so we must give up the demonstration by example.

It is necessary, then, to take all the consequences of our axioms and see whether they contain any contradiction. If the number of these consequences were finite, this would be easy; but their number is infinite—they are the whole of mathematics, or at least the whole of arithmetic.

What are we to do, then? Perhaps, if driven to it, we might repeat the reasoning of Section III. But, as I have said, *this reasoning is complete induction*, and it is precisely the principle of complete induction that we are engaged in justifying.

VI.

Hilbert's Logic.

I come now to Mr. Hilbert's important work, addressed to the Mathematical Congress at Heidelberg, a French translation of which, by M. Pierre Boutroux, appeared in *l'Enseignement Mathématique*, while an English translation by Mr. Halsted appeared in *The Monist*. In this work, in which we find the most profound thought, the author pursues an aim similar to Mr. Russell's, but he diverges on many points from his predecessor.

"However," he says, "if we look closely, we recognize that in logical principles, as they are commonly presented, certain arithmetical notions are found already implied; for instance, the notion of whole, and, to a certain extent, the notion of number. Thus we find ourselves caught in a circle, and that is why it seems to me necessary, if we wish to avoid all paradox, to develop the principles of logic and of arithmetic simultaneously."

We have seen above that what Mr. Hilbert says of the principles of logic, *as they are commonly presented*, applies equally to Mr. Russell's logic. For Mr. Russell logic is anterior to arithmetic, and for Mr. Hilbert they are "simultaneous." Further on we shall find other and yet deeper differences; but we will note them as they occur. I prefer to follow the development of Hilbert's thought step by step, quoting the more important passages verbatim.

"Let us first take into consideration the object 1." We notice that in acting thus we do not in any way imply the notion of number, for it is clearly understood that 1 here is nothing but a symbol, and that we do not in any way concern ourselves with knowing its signification. "The groups formed with this object, two, three, or several times repeated . . ." This time the case is quite altered, for if we introduce the words two, three, and, above all, several, we introduce the notion of number; and then the definition of the finite whole number that we find later on comes a trifle late. The author was much too wary not to perceive this *petitio principii*. And so, at the end of his work, he seeks to effect a real *patching-up*.

Hilbert then introduces two simple objects, 1 and =, and pictures all the combinations of these two objects, all the combinations of their combinations, and so on. It goes without saying that we must forget the ordinary signification of these two signs, and not attribute any to them. He then divides these combinations into two classes, that of entities and that of nonentities, and, until further orders, this partition is entirely arbitrary. Every affirmative proposition teaches us that a combination belongs to the class of entities, and every negative proposition teaches us that a certain combination belongs to the class ot nonentities.

VII.

We must now note a difference that is of the highest importance. For Mr. Russell a chance object, which he designates by x, is an absolutely indeterminate object, about which he assumes nothing. For Hilbert it is one of those combinations formed with the symbols 1 and $=$; he will not allow the introduction of anything but combinations of objects already defined. Moreover, Hilbert formulates his thought in the most concise manner, and I think I ought to reproduce his statement *in extenso:* "The indeterminates which figure in the axioms (in place of the 'some' or the 'all' of ordinary logic) represent exclusively the whole of the objects and combinations that we have already acquired in the actual state of the theory, or that we are in course of introducing. Therefore, when we deduce propositions from the axioms under consideration, it is these objects and these combinations alone that we have the right to substitute for the indeterminates. Neither must we forget that when we increase the number of the fundamental objects, the axioms at the same time acquire a new extension, and must, in consequence, be put to the proof afresh and, if necessary, modified."

The contrast with Mr. Russell's point of view is complete. According to this latter philosopher, we may substitute in place of x not only objects already known, but anything whatsoever. Russell is faithful to his point of view, which is that of comprehension. He starts with the general idea of entity, and enriches it more and more, even while he restricts it, by adding to it new qualities. Hilbert, on the contrary, only recognizes as possible entities combinations of objects already known; so that (looking only at one side of his thought) we might say that he takes the point of view of extension.

VIII.

Let us proceed with the exposition of Hilbert's ideas. He introduces two axioms which he enunciates in his symbolical language, but which signify, in the language of the uninitiated like us, that every quantity is equal to itself, and that every operation upon two identical quantities gives identical results. So stated they are evident, but such a presentation of them does not faithfully represent Hilbert's thought. For him mathematics has to combine only pure symbols, and a true mathematician must base his reasoning upon them without concerning himself with their meaning. Accordingly, his axioms are not for him what they are for the ordinary man.

He considers them as representing the definition by postulates of the symbol $=$, up to this time devoid of all signification. But in order to justify this definition, it is necessary to show that these two axioms do not lead to any contradiction.

For this purpose Hilbert makes use of the reasoning of Section III., without apparently perceiving that he is making complete induction.

IX.

The end of Mr. Hilbert's treatise is altogether enigmatical, and I will not dwell upon it. It is full of contradictions, and one feels that the author is vaguely conscious of the *petitio principii* he has been guilty of, and that he is vainly trying to plaster up the cracks in his reasoning.

What does this mean? It means that *when he comes to demonstrate that the definition of the whole number by the axiom of complete induction does not involve contradiction, Mr. Hilbert breaks down, just as Mr. Russell and M. Couturat broke down, because the difficulty is too great.*

X.

GEOMETRY.

Geometry, M. Couturat says, is a vast body of doctrine upon which complete induction does not intrude. This is true to a certain extent: we cannot say that it does not intrude at all, but that it intrudes very little. If we refer to Mr. Halsted's "Rational Geometry" (New York: John Wiley and Sons, 1904), founded on Hilbert's principles, we find the principle of induction intruding for the first time at page 114 (unless, indeed, I have not searched carefully enough, which is quite possible).

Thus geometry, which seemed, only a few years ago, the domain in which intuition held undisputed sway, is to-day the field in which the logisticians appear to triumph. Nothing could give a better measure of the importance of Hilbert's geometrical works, and of the profound impression they have left upon our conceptions.

But we must not deceive ourselves. *What is, in fact, the fundamental theorem of geometry? It is that the axioms of geometry do not involve contradiction, and this cannot be demonstrated without the principle of induction.*

How does Hilbert demonstrate this essential point? He does it by relying upon analysis, and, through it, upon arithmetic, and, through it, upon the principle of induction.

If another demonstration is ever discovered, it will still be necessary to rely on this principle, since the number of the possible consequences of the axioms which we have to show are not contradictory is infinite.

XI.

Conclusion.

Our conclusion is, first of all, that *the principle of induction cannot be regarded as the disguised definition of the whole number.*

Here are three truths:—

> The principle of complete induction;
> Euclid's postulate;
> The physical law by which phosphorus melts at 44° centigrade (quoted by M. Le Roy).

We say: these are three disguised definitions—the first that of the whole number, the second that of the straight line, and the third that of phosphorus.

I admit it for the second, but I do not admit it for the two others, and I must explain the reason of this apparent inconsistency.

In the first place, we have seen that a definition

is only acceptable if it is established that it does not involve contradiction. We have also shown that, in the case of the first definition, this demonstration is impossible; while in the case of the second, on the contrary, we have just recalled the fact that Hilbert has given a complete demonstration.

So far as the third is concerned, it is clear that it does not involve contradiction. But does this mean that this definition guarantees, as it should, the existence of the object defined? We are here no longer concerned with the mathematical sciences, but with the physical sciences, and the word existence has no longer the same meaning; it no longer signifies absence of contradiction, but objective existence.

This is one reason already for the distinction I make between the three cases, but there is a second. In the applications we have to make of these three notions, do they present themselves as defined by these three postulates?

The possible applications of the principle of induction are innumerable. Take, for instance, one of those we have expounded above, in which it is sought to establish that a collection of axioms cannot lead to a contradiction. For this purpose we consider one of the series of syllogisms that can be followed out, starting with these axioms as premises.

When we have completed the nth syllogism, we see that we can form still another, which will be the $(n+1)^{th}$: thus the number n serves for counting a series of successive operations; it is a number that can be obtained by successive additions. Accordingly, it is a number from which we can return to unity by *successive subtractions*. It is evident that we could not do so if we had $n = n - 1$, for then subtraction would always give us the same number. Thus, then, the way in which we have been brought to consider this number n involves a definition of the finite whole number, and this definition is as follows: *a finite whole number is that which can be obtained by successive additions, and which is such that n is not equal to $n - 1$.*

This being established, what do we proceed to do? We show that if no contradiction has occurred up to the nth syllogism, it will not occur any the more at the $(n+1)^{th}$, and we conclude that it will never occur. You say I have the right to conclude thus, because whole numbers are, by definition, those for which such reasoning is legitimate. But that involves another definition of the whole number, which is as follows: *a whole number is that about which we can reason by recurrence.* In the species it is that of which we can state that, if absence of contradiction at the moment of occurrence of a syllogism whose number is a whole number carries with it the absence of contradiction at the moment of occurrence of the syllogism whose number is the following whole number, then we need not fear any contradiction for any of the syllogisms whose numbers are whole numbers.

The two definitions are not identical. They are equivalent, no doubt, but they are so by virtue of an *a priori* synthetic judgment; we cannot pass from one to the other by purely logical processes. Consequently, we have no right to adopt the second after having introduced the whole number by a road which presupposes the first.

On the contrary, what happens in the case of the

straight line? I have already explained this so often that I feel some hesitation about repeating myself once more. I will content myself with a brief summary of my thought.

We have not, as in the previous case, two equivalent definitions logically irreducible one to the other. We have only one expressible in words. It may be said that there is another that we feel without being able to enunciate it, because we have the intuition of a straight line, or because we can picture a straight line. But, in the first place, we cannot picture it in geometric space, but only in representative space; and then we can equally well picture objects which possess the other properties of a straight line, and not that of satisfying Euclid's postulate. These objects are "non-Euclidian straight lines," which, from a certain point of view, are not entities destitute of meaning, but circles (true circles of true space) orthogonal to a certain sphere. If, among these objects equally susceptible of being pictured, it is the former (the Euclidian straight lines) that we call straight lines, and not the latter (the non-Euclidian straight lines), it is certainly so by definition.

And if we come at last to the third example, the definition of phosphorus, we see that the true definition would be: phosphorus is this piece of matter that I see before me in this bottle.

XII.

Since I am on the subject, let me say one word more. Concerning the example of phosphorus, I said: "This proposition is a true physical law that can be verified, for it means: all bodies which possess all the properties of phosphorus except its melting-point, melt, as it does, at 44° centigrade." It has been objected that this law is not verifiable, for if we came to verify that two bodies resembling phosphorus melt one at 44° and the other at 50° centigrade, we could always say that there is, no doubt, besides the melting-point, some other property in which they differ.

This was not exactly what I meant to say, and I should have written: "all bodies which possess such and such properties in finite number (namely, the properties of phosphorus given in chemistry books, with the exception of its melting-point) melt at 44° centigrade."

In order to make still clearer the difference between the case of the straight line and that of phosphorus, I will make one more remark. The straight line has several more or less imperfect images in nature, the chief of which are rays of light and the axis of rotation of a solid body. Assuming that we ascertain that the ray of light does not satisfy Euclid's postulate (by showing, for instance, that a star has a negative parallax), what shall we do? Shall we conclude that, as a straight line is by definition the trajectory of light, it does not satisfy the definition, or, on the contrary, that, as a straight line by definition satisfies the postulate, the ray of light is not rectilineal?

Certainly we are free to adopt either definition, and, consequently, either conclusion. But it would be foolish to adopt the former, because the ray of light probably satisfies in a most imperfect way not only Euclid's postulate but the other properties of the straight line; because, while it deviates from the Euclidian straight, it deviates none the less from the axis of rotation of solid bodies, which is another imperfect image of the straight line; and lastly, because it is, no doubt, subject to change, so that such and such a line which was straight yesterday will no longer be so to-morrow if some physical circumstance has altered.

Assume, now, that we succeed in discovering that phosphorus melts not at 44° but at 43·9° centigrade. Shall we conclude that, as phosphorus is by definition that which melts at 44°, this substance that we called phosphorus is not true phosphorus, or, on the contrary, that phosphorus melts at 43·9°? Here, again, we are free to adopt either definition, and, consequently, either conclusion; but it would be foolish to adopt the former, because we cannot change the name of a substance every time we add a fresh decimal to its melting-point.

XIII.

To sum up, Mr. Russell and Mr. Hilbert have both made a great effort, and have both of them written a book full of views that are original, profound, and often very true. These two books furnish us with subject for much thought, and there is much that we can learn from them. Not a few of their results are substantial and destined to survive.

But to say that they have definitely settled the controversy between Kant and Leibnitz and destroyed the Kantian theory of mathematics is evidently untrue. I do not know whether they actually imagined they had done it, but if they did they were mistaken.

V.

THE LAST EFFORTS OF THE LOGISTICIANS.

I.

THE logisticians have attempted to answer the foregoing considerations. For this purpose they have been obliged to transform logistic, and Mr. Russell in particular has modified his original views on certain points. Without entering into the details of the controversy, I should like to return to what are, in my opinion, the two most important questions. Have the rules of logistic given any proof of fruitfulness and of infallibility? Is it true that they make it possible to demonstrate the principle of complete induction without any appeal to intuition?

II.

THE INFALLIBILITY OF LOGISTIC.

As regards fruitfulness, it seems that M. Couturat has most childish illusions. Logistic, according to him, lends "stilts and wings" to discovery, and on the following page he says, "*It is ten years* since Signor Peano published the first edition of his "Formulaire."

What! you have had wings for ten years, and you haven't flown yet!

I have the greatest esteem for Signor Peano, who

has done some very fine things (for instance, his curve which fills a whole area); but, after all, he has not gone any farther, or higher, or faster than the majority of wingless mathematicians, and he could have done everything just as well on his feet.

On the contrary, I find nothing in logistic for the discoverer but shackles. It does not help us at all in the direction of conciseness, far from it; and if it requires 27 equations to establish that 1 is a number, how many will it require to demonstrate a real theorem? If we distinguish, as Mr. Whitehead does, the individual x, the class whose only member is x, which we call ιx, then the class whose only member is the class whose only member is x, which we call $\iota\iota x$, do we imagine that these distinctions, however useful they may be, will greatly expedite our progress?

Logistic forces us to say all that we commonly assume, it forces us to advance step by step; it is perhaps surer, but it is not more expeditious.

It is not wings you have given us, but leading-strings. But we have the right to demand that these leading-strings should keep us from falling; this is their only excuse. When an investment does not pay a high rate of interest, it must at least be a gilt-edged security.

Must we follow your rules blindly? Certainly, for otherwise it would be intuition alone that would enable us to distinguish between them. But in that case they must be infallible, for it is only in an infallible authority that we can have blind confidence. Accordingly, this is a necessity for you: you must be infallible or cease to exist.

You have no right to say to us: "We make mistakes,

it is true, but you make mistakes too." For us, making mistakes is a misfortune, a very great misfortune, but for you it is death.

Neither must you say, "Does the infallibility of arithmetic prevent errors of addition?" The rules of calculation are infallible, and yet we find people making mistakes *through not applying these rules.* But a revision of their calculation will show at once just where they went astray. Here the case is quite different. The logisticians *have applied* their rules, and yet they have fallen into contradiction. So true is this, that they are preparing to alter these rules and "sacrifice the notion of class." Why alter them if they were infallible?

"We are not obliged," you say, "to solve *hic et nunc* all possible problems." Oh, we do not ask as much as that. If, in face of a problem, you gave *no* solution, we should have nothing to say; but, on the contrary, you give *two,* and these two are contradictory, and consequently one at least of them is false, and it is this that constitutes a failure.

Mr. Russell attempts to reconcile these contradictions, which can only be done, according to him, "by restricting or even sacrificing the notion of class." And M. Couturat, discounting the success of this attempt, adds: "If logisticians succeed where others have failed, M. Poincaré will surely recollect this sentence, and give logistic the credit of the solution."

Certainly not. Logistic exists; it has its code, which has already gone through four editions; or, rather, it is this code which is logistic itself. Is Mr. Russell preparing to show that one at least of the two contradictory arguments has transgressed the code? Not in the very least; he is preparing to alter these laws and to revoke a certain number of them. If he succeeds, I shall give credit to Mr. Russell's intuition, and not to Peanian Logistic, which he will have destroyed.

III.

Liberty of Contradiction.

I offered two principal objections to the definition of the whole number adopted by the logisticians. What is M. Couturat's answer to the first of these objections?

What is the meaning in mathematics of the word *to exist?* It means, I said, to be free from contradiction. This is what M. Couturat disputes. "Logical existence," he says, "is quite a different thing from absence of contradiction. It consists in the fact that a class is not empty. To say that some *a*'s exist is, by definition, to assert that the class *a* is not void." And, no doubt, to assert that the class *a* is not void is, by definition, to assert that some *a*'s exist. But one of these assertions is just as destitute of meaning as the other if they do not both signify either that we can see or touch *a*, which is the meaning given them by physicists or naturalists, or else that we can conceive of an *a* without being involved in contradictions, which is the meaning given them by logicians and mathematicians.

In M. Couturat's opinion it is not non-contradiction that proves existence, but existence that proves non-contradiction. In order to establish the existence of a class, we must accordingly establish, by an *example*, that there is an individual belonging to that class.

"But it will be said, How do we demonstrate the existence of this individual? Is it not necessary that this existence should be established, to enable us to deduce the existence of the class of which it forms part? It is not so. Paradoxical as the assertion may appear, we never demonstrate the existence of an individual. Individuals, from the very fact that they are individuals, are always considered as existing. We have never to declare that an individual exists, absolutely speaking, but only that it exists in a class." M. Couturat finds his own assertion paradoxical, and he will certainly not be alone in so finding it. Nevertheless it must have some sense, and it means, no doubt, that the existence of an individual alone in the world, of which nothing is asserted, cannot involve contradiction. As long as it is quite alone, it is evident that it cannot interfere with any one. Well, be it so; we will admit the existence of the individual, "absolutely speaking," but with it we have nothing to do. It still remains to demonstrate the existence of the individual "in a class," and, in order to do this, you will still have to prove that the assertion that such an individual belongs to such a class is neither contradictory in itself nor with the other postulates adopted.

"Accordingly," M. Couturat continues, "to assert that a definition is not valid unless it is first proved that it is not contradictory, is to impose an arbitrary and improper condition." The claim for the liberty of contradiction could not be stated in more emphatic or haughtier terms. "In any case, the *onus probandi* rests with those who think these principles are contradictory." Postulates are presumed to be compatible, just as a prisoner is presumed to be innocent, until the contrary is proved.

It is unnecessary to add that I do not acquiesce in this claim. But, you say, the demonstration you demand of us is impossible, and you cannot require us to "aim at the moon." Excuse me; it is impossible for you, but not for us who admit the principle of induction as an *a priori* synthetic judgment. This would be necessary for you as it is for us.

In order to demonstrate that a system of postulates does not invclve contradiction, it is necessary to apply the principle of complete induction. Not only is there nothing "extraordinary" in this method of reasoning, but it is the only correct one. It is not "inconceivable" that any one should ever have used it, and it is not difficult to find "examples and precedents." In my article I have quoted two, and they were borrowed from Hilbert's pamphlet. He is not alone in having made use of it, and those who have not done so have been wrong. What I reproach Hilbert with, is not that he has had recourse to it (a born mathematician such as he could not but see that a demonstration is required, and that this is the only possible one), but that he has had recourse to it without recognizing the reasoning by recurrence.

IV.

The Second Objection.

I had noted a second error of the logisticians in Hilbert's article. To-day Hilbert is excommunicated, and M. Couturat no longer considers him as a logistician. He will therefore, ask me if I have found the same mistake in the orthodox logisticians. I have not seen it in the pages I have read, but I do not know whether I should find it in the three hundred pages they have written that I have no wish to read.

Only, they will have to commit the error as soon as they attempt to make any sort of an application of mathematical science. The eternal contemplation of its own navel is not the sole object of this science. It touches nature, and one day or other it will come into contact with it. Then it will be necessary to shake off purely verbal definitions and no longer to content ourselves with words.

Let us return to Mr. Hilbert's example. It is still a question of reasoning by recurrence and of knowing whether a system of postulates is not contradictory. M. Couturat will no doubt tell me that in that case it does not concern him, but it may perhaps interest those who do not claim, as he does, the liberty of contradiction.

We wish to establish, as above, that we shall not meet with contradiction after some particular number of arguments, a number which may be as large as you please, provided it is finite. For this purpose we must apply the principle of induction. Are we to understand here by finite number every number to which the principle of induction applies? Evidently not, for otherwise we should be involved in the most awkward consequences.

To have the right to lay down a system of postulates, we must be assured that they are not contradictory. This is a truth that is admitted by *the majority* of scientists; I should have said *all* before reading M. Couturat's last article. But what does it signify? Does it mean that we must be sure of not meeting with contradiction after a *finite* number of propositions, the *finite* number being, by definition, that which possesses all the properties of a recurrent nature in such a way that if one of these properties were found wanting—if, for instance, we came upon a contradiction—we should *agree* to say that the number in question was not finite?

In other words, do we mean that we must be sure of not meeting a contradiction, with this condition, that we agree to stop just at the moment when we are on the point of meeting one? The mere statement of such a proposition is its sufficient condemnation.

Thus not only does Mr. Hilbert's reasoning assume the principle of induction, but he assumes that this principle is given us, not as a simple definition, but as an *a priori* synthetic judgment.

I would sum up as follows:—

A demonstration is necessary.

The only possible demonstration is the demonstration by recurrence.

This demonstration is legitimate only if the principle of induction is admitted, and if it is regarded not as a definition but as a synthetic judgment.

V.

The Cantorian Antinomies.

I will now take up the examination of Mr. Russell's new treatise. This treatise was written with the object of overcoming the difficulties raised by those *Cantorian antinomies* to which I have already made frequent allusion. Cantor thought it possible to construct a Science of the Infinite. Others have advanced further along the path he had opened, but they very soon ran against strange contradictions. These antinomies are already numerous, but the most celebrated are:—

1. Burali-Forti's antinomy.
2. The Zermelo-König antinomy.
3. Richard's antinomy.

Cantor had demonstrated that ordinal numbers (it is a question of transfinite ordinal numbers, a new notion introduced by him) can be arranged in a lineal series; that is to say, that of two unequal ordinal numbers, there is always one that is smaller than the other. Burali-Forti demonstrates the contrary; and indeed, as he says in substance, if we could arrange *all* the ordinal numbers in a lineal series, this series would define an ordinal number that would be greater than *all* the others, to which we could then add 1 and so obtain yet another ordinal number which would be still greater. And this is contradictory.

We will return later to the Zermelo-König antinomy, which is of a somewhat different nature. Richard's antinomy is as follows (*Revue générale des Sciences*, June 30, 1905). Let us consider all the decimal numbers that can be defined with the help of a finite number of words. These decimal numbers form an aggregate E, and it is easy to see that this aggregate is denumerable—that is to say, that it is possible to *number* the decimal numbers of this aggregate from one to infinity. Suppose the numeration effected, and let us define a number N in the following manner. If the nth decimal of the nth number of the aggregate E is

0, 1, 2, 3, 4, 5, 6, 7, 8, or 9,

the nth decimal of N will be

1, 2, 3, 4, 5, 6, 7, 8, 1, or 1.

As we see, N is not equal to the nth number of E, and since n is any chance number, N does not belong to E, and yet N should belong to this aggregate, since we have defined it in a finite number of words.

We shall see further on that M. Richard himself has, with much acuteness, given the explanation of his paradox, and that his explanation can be extended, *mutatis mutandis*, to the other paradoxes of like nature. Mr. Russell quotes another rather amusing antinomy:

What is the smallest whole number that cannot be defined in a sentence formed of less than a hundred English words?

This number exists, and, indeed, the number of numbers capable of being defined by such a sentence is evidently finite, since the number of words in the English language is not infinite. Therefore among them there will be one that is smaller than all the others.

On the other hand the number does not exist, for its definition involves contradiction. The number, in fact, is found to be defined by the sentence in italics, which is formed of less than a hundred English words, and, by definition, the number must not be capable of being defined by such a sentence.

VI.

Zigzag Theory and No Classes Theory.

What is Mr. Russell's attitude in face of these contradictions? After analysing those I have just spoken of, and quoting others, after putting them in a form that recalls Epimenides, he does not hesitate to conclude as follows:—

"A propositional function of one variable does not always determine a class."* A "propositional function" (that is to say, a definition) or "norm" can be "non-predicative." And this does not mean that these non-predicative propositions determine a class that is empty or void; it does not mean that there is no value of x that satisfies the definition and can be one of the elements of the class. The elements exist, but they have no right to be grouped together to form a class.

But this is only the beginning, and we must know how to recognize whether a definition is or is not predicative. For the purpose of solving this problem, Mr. Russell hesitates between three theories, which he calls—

A. The zigzag theory.
B. The theory of limitation of size.
C. The no classes theory.

According to the zigzag theory, "definitions (propositional functions) determine a class when they are fairly simple, and only fail to do so when they are complicated and recondite." Now who is to decide

* This and the following quotations are from Mr. Russell's paper, "On some difficulties in the theory of transfinite numbers and order types," *Proceedings of the London Mathematical Society. Ser. 2, Vol. 4, Part 1.*

whether a definition can be regarded as sufficiently simple to be acceptable? To this question we get no answer except a candid confession of powerlessness. "The axioms as to what functions are predicative have to be exceedingly complicated, and cannot be recommended by any intrinsic plausibility. This is a defect which might be remedied by greater ingenuity, or by the help of some hitherto unnoticed distinction. But hitherto, in attempting to set up axioms for this theory, I have found no guiding principle except the avoidance of contradictions."

This theory therefore remains very obscure. In the darkness there is a single glimmer, and that is the word zigzag. What Mr. Russell calls *zigzagginess* is no doubt this special character which distinguishes the argument of Epimenides.

According to the theory of limitation of size, a class must not be too extensive. It may, perhaps, be infinite, but it must not be too infinite.

But we still come to the same difficulty. At what precise moment will it begin to be too extensive? Of course this difficulty is not solved, and Mr. Russell passes to the third theory.

In the no classes theory all mention of the word *class* is prohibited, and the word has to be replaced by various periphrases. What a change for the logisticians who speak of nothing but class and classes of classes! The whole of Logistic will have to be refashioned. Can we imagine the appearance of a page of Logistic when all propositions dealing with class have been suppressed? There will be nothing left but a few scattered survivors in the midst of a blank page. *Apparent rari nantes in gurgite vasto.*

However that may be, we understand Mr. Russell's hesitation at the modifications to which he is about to submit the fundamental principles he has hitherto adopted. Criteria will be necessary to decide whether a definition is too complicated or too extensive, and these criteria cannot be justified except by an appeal to intuition.

It is towards the no classes theory that Mr. Russell eventually inclines.

However it be, Logistic must be refashioned, and it is not yet known how much of it can be saved. It is unnecessary to add that it is Cantorism and Logistic alone that are in question. The true mathematics, the mathematics that is of some use, may continue to develop according to its own principles, taking no heed of the tempests that rage without, and step by step it will pursue its wonted conquests, which are decisive and have never to be abandoned.

VII.

The True Solution.

How are we to choose between these different theories? It seems to me that the solution is contained in M. Richard's letter mentioned above, which will be found in the *Revue Générale des Sciences* of June 30, 1905. After stating the antinomy that I have called Richard's antinomy, he gives the explanation.

Let us refer to what was said of this antinomy in Section V. E is the aggregate of *all* the numbers that can be defined by a finite number of words, *without introducing the notion of the aggregate E itself*, otherwise the definition of E would contain a vicious circle, for we cannot define E by the aggregate E itself.

Now we have defined N by a finite number of words, it is true, but only with the help of the notion of the aggregate E, and that is the reason why N does not form a part of E.

In the example chosen by M. Richard, the conclusion is presented with complete evidence, and the evidence becomes the more apparent on a reference to the actual text of the letter. But the same explanation serves for the other antinomies, as may be easily verified.

Thus *the definitions that must be regarded as non-predicative are those which contain a vicious circle.* The above examples show sufficiently clearly what I mean by this. Is this what Mr. Russell calls "zigzagginess"? I merely ask the question without answering it.

VIII.

THE DEMONSTRATIONS OF THE PRINCIPLE OF INDUCTION.

We will now examine the so-called demonstrations of the principle of induction, and more particularly those of Mr. Whitehead and Signor Burali-Forti.

And first we will speak of Whitehead's, availing ourselves of some new denominations happily introduced by Mr. Russell in his recent treatise.

We will call *recurrent class* every class of numbers that includes zero, and also includes $n+1$ if it includes n.

We will call *inductive number* every number which forms a part of *all* recurrent classes.

Upon what condition will this latter definition, which plays an essential part in Whitehead's demonstration, be "predicative" and consequently acceptable?

Following upon what has been said above, we must understand by *all* recurrent classes all those whose definition does not contain the notion of inductive number; otherwise we shall be involved in the vicious circle which engendered the antinomies.

Now, *Whitehead has not taken this precaution.*

Whitehead's argument is therefore vicious; it is the same that led to the antinomies. It was illegitimate when it gave untrue results, and it remains illegitimate when it leads by chance to a true result.

A definition which contains a vicious circle defines nothing. It is of no use to say we are sure, whatever be the meaning given to our definition, that there is at least zero which belongs to the class of inductive numbers. It is not a question of knowing whether this class is empty, but whether it can be rigidly delimited. A "non-predicative class" is not an empty class, but a class with uncertain boundaries.

It is unnecessary to add that this particular objection does not invalidate the general objections that apply to all the demonstrations.

IX.

Signor Burali-Forti has given another demonstration in his article "Le Classi finite" (*Atti di Torino*, Vol. xxxii). But he is obliged to admit two postulates:

The first is that there exists always at least one infinite class.

The second is stated thus:—

$$u \,\epsilon\, K(K - \iota\Lambda) \,.\supset.\, u < v'u.$$

The first postulate is no more evident than the principle to be demonstrated. The second is not only not evident, but it is untrue, as Mr. Whitehead has shown, as, moreover, the veriest schoolboy could have seen at the first glance if the axiom had been stated in intelligible language, since it means: the number of combinations that can be formed with several objects is smaller than the number of those objects.

X.

Zermelo's Axiom.

In a celebrated demonstration, Signor Zermelo relies on the following axiom:

In an aggregate of any kind (or even in each of the aggregates of an aggregate of aggregates) we can always select one element *at random* (even if the aggregate of aggregates contains an infinity of aggregates).

This axiom had been applied a thousand times without being stated, but as soon as it was stated, it raised doubts. Some mathematicians, like M. Borel, rejected it resolutely, while others admitted it. Let us see what Mr. Russell thinks of it according to his last article.

He pronounces no opinion, but the considerations which he gives are most suggestive.

To begin with a picturesque example, suppose that we have as many pairs of boots as there are whole numbers, so that we can number *the pairs* from 1 to infinity, how many boots shall we have? Will the number of boots be equal to the number of pairs? It will be so if, in each pair, the right boot is distinguishable from the left; it will be sufficient in fact to give the number $2n-1$ to the right boot of the nth pair, and the number $2n$ to the left boot of the nth pair. But it will not be so if the right boot is similar to the left, because such an operation then becomes impossible; unless we admit Zermelo's axiom, since in that case we can select *at random* from each pair the boot we regard as the right.

XI.

CONCLUSIONS.

A demonstration really based upon the principles of Analytical Logic will be composed of a succession of propositions; some, which will serve as premises, will be identities or definitions; others will be deduced from the former step by step; but although the connexion between each proposition and the succeeding proposition can be grasped immediately, it is not obvious at a glance how it has been possible to pass from the first to the last, which we may be tempted to look upon as a new truth. But if we replace successively the various expressions that are used by their definitions, and if we pursue this operation to the furthest possible limit, there will be nothing left at the end but identities, so that all will be reduced to one immense tautology. Logic therefore remains barren, unless it is fertilized by intuition.

This is what I wrote formerly. The logisticians assert the contrary, and imagine that they have proved it by effectively demonstrating new truths. But what mechanism have they used?

Why is it that by applying to their arguments the procedure I have just described, that is, by replacing the terms defined by their definitions, we do not see them melt into identities like the ordinary arguments? It is because the procedure is not applicable to them. And why is this? Because their definitions are non-predicative and present that kind of hidden vicious circle I have pointed out above, and non-predicative definitions cannot be substituted for the term defined. Under these conditions, *Logistic is no longer barren, it engenders antinomies.*

It is the belief in the existence of actual infinity that has given birth to these non-predicative definitions. I must explain myself. In these definitions we find the word *all*, as we saw in the examples quoted above. The word *all* has a very precise meaning when it is a question of a finite * number of objects; but for it still to have a precise meaning when the number of the objects is infinite, it is necessary that there should exist an actual infinity. Otherwise *all* these objects cannot be conceived as existing prior to their definition, and then, if the definition of a notion N depends on *all* the objects A, it may be tainted with the vicious circle, if among the objects A there is one that cannot be defined without bringing in the notion N itself.

The rules of formal logic simply express the properties of all the possible classifications. But in order that they should be applicable, it is necessary that these classifications should be immutable and not require to be modified in the course of the argument. If we have only to classify a finite number of objects, it is easy to preserve these classifications without

* The original has "infinite," obviously a slip.

change. If the number of the objects is indefinite, that is to say if we are constantly liable to find new and unforeseen objects springing up, it may happen that the appearance of a new object will oblige us to modify the classification, and it is thus that we are exposed to the antinomies.

There is no actual infinity. The Cantorians forgot this, and so fell into contradiction. It is true that Cantorism has been useful, but that was when it was applied to a real problem, whose terms were clearly defined, and then it was possible to advance without danger.

Like the Cantorians, the logisticians have forgotten the fact, and they have met with the same difficulties. But it is a question whether they took this path by accident or whether it was a necessity for them.

In my view, there is no doubt about the matter; belief in an actual infinity is essential in the Russellian logistic, and this is exactly what distinguishes it from the Hilbertian logistic. Hilbert takes the point of view of extension precisely in order to avoid the Cantorian antinomies. Russell takes the point of view of comprehension, and consequently for him the genus is prior to the species, and the *summum genus* prior to all. This would involve no difficulty if the *summum genus* were finite; but if it is infinite, it is necessary to place the infinite before the finite—that is to say, to regard the infinite as actual.

And we have not only infinite classes; when we pass from the genus to the species by restricting the concept by new conditions, the number of these conditions is still infinite, for they generally express that the object under consideration is in such and such a relation with all the objects of an infinite class.

But all this is ancient history. Mr. Russell has realized the danger and is going to reconsider the matter. He is going to change everything, and we must understand clearly that he is preparing not only to introduce new principles which permit of operations formerly prohibited, but also to prohibit operations which he formerly considered legitimate. He is not content with adoring what he once burnt, but he is going to burn what he once adored, which is more serious. He is not adding a new wing to the building, but sapping its foundations.

The old Logistic is dead, and so true is this, that the zigzag theory and the no classes theory are already disputing the succession. We will wait until the new exists before we attempt to judge it.

BOOK III.

THE NEW MECHANICS.

I.

MECHANICS AND RADIUM.

I.

INTRODUCTION.

ARE the general principles of Dynamics, which have served since Newton's day as the foundation of Physical Science, and appear immutable, on the point of being abandoned, or, at the very least, profoundly modified? This is the question many people have been asking for the last few years. According to them the discovery of radium has upset what were considered the most firmly rooted scientific doctrines, the impossibility of the transmutation of metals on the one hand, and, on the other, the fundamental postulates of Mechanics. Perhaps they have been in too great haste to consider these novelties as definitely established, and to shatter our idols of yesterday; perhaps it would be well to await more numerous and more convincing experiments. It is none the less necessary that we should at once acquire a knowledge of the new doctrines and of the arguments, already most weighty, upon which they rely.

I will first recall in a few words what these principles are.

A. The motion of a material point, isolated and unaffected by any exterior force, is rectilineal and uniform. This is the principle of inertia; no acceleration without force.

B. The acceleration of a moving point has the same direction as the resultant of all the forces to which the point is subjected; it is equal to the quotient of this resultant by a coëfficient called the *mass* of the moving point.

The mass of a moving point, thus defined, is constant; it does not depend upon the velocity acquired by the point, it is the same whether the force is parallel to this velocity and only tends to accelerate or retard the motion of the point, or whether it is, on the contrary, perpendicular to that velocity and tends to cause the motion to deviate to right or left, that is to say to *curve* the trajectory.

C. All the forces to which a material point is subjected arise from the action of other material points; they depend only upon the *relative* positions and velocities of these different material points.

By combining the two principles B and C we arrive at the *principle of relative motion*, by virtue of which the laws of motion of a system are the same whether we refer the system to fixed axes, or whether we refer it to moving axes animated with a rectilineal and uniform forward motion, so that it is impossible to distinguish absolute motion from a relative motion referred to such moving axes.

D. If a material point A acts upon another material point B, the body B reacts upon A, and these two actions are two forces that are equal and directly opposite to one another. This is *the principle of the equality of action and reaction*, or more briefly, *the principle of reaction*.

Astronomical observations, and the commonest physical phenomena, seem to have afforded the most complete, unvarying, and precise confirmation of these principles. That is true, they tell us now, but only because we have never dealt with any but low velocities. Mercury, for instance, which moves faster than any of the other planets, scarcely travels sixty miles a second—Would it behave in the same way if it travelled a thousand times as fast? It is clear that we have still no cause for anxiety; whatever may be the progress of automobilism, it will be some time yet before we have to give up applying the classical principles of Dynamics to our machines.

How is it then that we have succeeded in realizing velocities a thousand times greater than that of Mercury, equal, for instance, to a tenth or a third of the velocity of light, or coming nearer to it even than that? It is by the help of the cathode rays and the rays of radium.

We know that radium emits three kinds of rays, which are designated by the three Greek letters α, β, γ. In what follows, unless I specifically state the contrary, I shall always speak of the β rays, which are analogous to the cathode rays.

After the discovery of the cathode rays, two opposite theories were propounded. Crookes attributed the phenomena to an actual molecular bombardment, Hertz to peculiar undulations of the ether. It was a repetition of the controversy that had divided physicists a century before with regard to light. Crookes returned to the emission theory, abandoned in the case of light, while Hertz held to the undulatory theory. The facts seemed to be in favour of Crookes.

It was recognized in the first place that the cathode rays carry with them a negative electric charge: they are deviated by a magnetic and by an electric field, and these deviations are precisely what would be produced by these same fields upon projectiles animated with a very great velocity, and highly charged with negative electricity. These two deviations depend upon two quantities; the velocity on the one hand, and the proportion of the projectile's electric charge to its mass on the other. We cannot know the absolute value of this mass, nor that of the charge, but only their proportion. It is clear in fact, that if we double both the charge and the mass, without changing the velocity, we shall double the force that tends to deviate the projectile; but as its mass is similarly doubled, the observable acceleration and deviation will not be changed. Observation of the two deviations will accordingly furnish us with two equations for determining these two unknown quantities. We find a velocity of 6,000 to 20,000 miles a second. As for the proportion of the charge to the mass, it is very great; it may be compared with the corresponding proportion in the case of a hydrogen ion in electrolysis, and we find then that a cathode projectile carries with it about a thousand times as much electricity as an equal mass of hydrogen in an electrolyte.

In order to confirm these views, we should require a direct measure of this velocity, that could then be compared with the velocity so calculated. Some old experiments of Sir J. J. Thomson's had given results more than a hundred times too low, but they were subject to certain causes of error. The question has been taken up again by Wiechert, with the help of an arrangement by which he makes use of the Hertzian oscillations, and this has given results in accordance with the theory, at least in the matter of magnitude, and it would be most interesting to take up these experiments again. However it be, the theory of undulations seems to be incapable of accounting for this body of facts.

The same calculations made upon the β rays of radium have yielded still higher velocities—60,000, 120,000 miles a second, and even more. These velocities greatly surpass any that we know. It is true that light, as we have long known, travels 186,000 miles a second, but it is not a transportation of matter, while, if we adopt the emission theory for the cathode rays, we have material molecules actually animated with the velocities in question, and we have to enquire whether the ordinary laws of Mechanics are still applicable to them.

II.

Longitudinal and Transversal Mass.

We know that electric currents give rise to phenomena of induction, in particular to *self-induction.* When a current increases it develops an electro-motive force of self-induction which tends to oppose the current. On the contrary, when the current decreases, the electro-motive force of self-induction tends to maintain the current. Self-induction then opposes all variation in the intensity of a current, just as in

Mechanics, the inertia of a body opposes all variation in its velocity. *Self-induction is an actual inertia.* Everything takes place as if the current could not be set up without setting the surrounding ether in motion, and as if the inertia of this ether consequently tended to keep the intensity of the current constant. The inertia must be overcome to set up the current, and it must be overcome again to make it cease.

A cathode ray, which is a rain of projectiles charged with negative electricity, can be likened to a current. No doubt this current differs, at first sight at any rate, from the ordinary conduction currents, where the matter is motionless and the electricity circulates through the matter. It is a *convection current*, where the electricity is attached to a material vehicle and carried by the movement of that vehicle. But Rowland has proved that convection currents produce the same magnetic effects as conduction currents. They must also produce the same effects of induction. Firstly, if it were not so, the principle of the conservation of energy would be violated; and secondly, Crémien and Pender have employed a method in which these effects of induction are *directly* demonstrated.

If the velocity of a cathode corpuscle happens to vary, the intensity of the corresponding current will vary equally, and there will be developed effects of self-induction which tend to oppose this variation. These corpuscles must therefore possess a double inertia, first their actual inertia, and then an apparent inertia due to self-induction, which produces the same effects. They will therefore have a total apparent mass, composed of their real mass and of a fictitious mass of electro-magnetic origin. Calculation shows that this fictitious mass varies with the velocity (when this is comparable with the velocity of light), and that the force of the inertia of self-induction is not the same when the velocity of the projectile is increased or diminished, as when its direction is changed, and accordingly the same holds good of the apparent total force of inertia.

The total apparent mass is therefore not the same when the actual force applied to the corpuscle is parallel with its velocity and tends to accelerate its movement, as when it is perpendicular to the velocity and tends to alter its direction. Accordingly we must distinguish between the *total longitudinal mass* and the *total transversal mass*, and, moreover, these two total masses depend upon the velocity. Such are the results of Abraham's theoretical work.

In the measurements spoken of in the last section, what was it that was determined by measuring the two deviations? The velocity on the one hand, and on the other the proportion of the charge to the *total transversal mass.* Under these conditions, how are we to determine what are the proportions, in this total mass, of the actual mass and of the fictitious electro-magnetic mass? If we had only the cathode rays properly so called, we could not dream of doing so, but fortunately we have the rays of radium, whose velocity, as we have seen, is considerably higher. These rays are not all identical, and do not behave in the same way under the action of an electric and a magnetic field. We find that the electric deviation is a function of the magnetic deviation, and by receiving upon a sensitive plate rays of radium that have been subjected to the action of the two fields, we can photograph the curve which represents the relation between these two deviations. This is what Kaufmann has done, and he has deduced the relation between the velocity and the proportion of the charge to the total apparent mass, a proportion that we call ϵ.

We might suppose that there exist several kinds of rays, each characterized by a particular velocity, by a particular charge, and by a particular mass; but this hypothesis is most improbable. What reason indeed could there be why all the corpuscles of the same mass should always have the same velocity? It is more natural to suppose that the charge and the *actual* mass are the same for all the projectiles, and that they differ only in velocity. If the proportion ϵ is a function of the velocity, it is not because the actual mass varies with the velocity, but, as the fictitious electro-magnetic mass depends upon that velocity, the total apparent mass, which is alone observable, must depend upon it also, even though the actual mass does not depend upon it but is constant.

Abraham's calculations make us acquainted with the law in accordance with which the *fictitious* mass varies as a function of the velocity, and Kaufmann's experiment makes us acquainted with the law of variation of the *total* mass. A comparison of these two laws will therefore enable us to determine the proportion of the *actual* mass to the total mass.

Such is the method employed by Kaufmann to determine this proportion. The result is most surprising: *the actual mass is nil.*

We have thus been led to quite unexpected conceptions. What had been proved only in the case of the cathode corpuscles has been extended to all bodies. What we call mass would seem to be nothing but an appearance, and all inertia to be of electromagnetic origin. But if this be true, mass is no longer constant; it increases with the velocity: while apparently constant for velocities up to as much as 600 miles a second, it grows thenceforward and becomes infinite for the velocity of light. Transversal mass is no longer equal to longitudinal mass, but only about equal if the velocity is not too great. Principle B of mechanics is no longer true.

III.

CANAL-RAYS.

At the point we have reached, this conclusion may seem premature. Can we apply to the whole of matter what has only been established for these very light corpuscles which are only an emanation of matter and perhaps not true matter? But before broaching this question, we must say a word about another kind of rays—I mean the *canal-rays*, Goldstein's *Kanalstrahlen*. Simultaneously with the cathode rays charged with negative electricity, the cathode emits canal-rays charged with positive electricity. In general these canal-rays, not being repelled by the cathode, remain confined in the immediate neighbourhood of that cathode, where they form the "buff stratum" that is not very easy to detect. But if the cathode is pierced with holes and blocks the tube almost completely, the canal-rays will be generated *behind* the cathode, in the opposite direction from that of the cathode rays, and it will become possible to study them. It is thus that we have been enabled to demonstrate their positive charge and to show that the magnetic and electric deviations still exist, as in the case of the cathode rays, though they are much weaker.

Radium likewise emits rays similar to the canal-rays, and relatively very absorbable, which are called α rays.

As in the case of the cathode rays, we can measure the two deviations and deduce the velocity and the proportion ϵ. The results are less constant than in the case of the cathode rays, but the velocity is lower, as is also the proportion ϵ. The positive corpuscles are less highly charged than the negative corpuscles; or if, as is more natural, we suppose that the charges are equal and of opposite sign, the positive corpuscles are much larger. These corpuscles, charged some positively and others negatively, have been given the name of *electrons*.*

IV.

Lorentz's Theory.

But the electrons do not only give evidence of their existence in these rays in which they appear

* The name is now applied only to the negative corpuscles, which seem to possess no actual mass and only a fictitious electro-magnetic mass, and not to the canal-rays, which appear to consist of ordinary chemical atoms positively charged, owing to the fact that they have lost one or more of the electrons they possess in their ordinary neutral state.

to us animated with enormous velocities. We shall see them in very different parts, and it is they that explain for us the principal phenomena of optics and of electricity. The brilliant synthesis about which I am going to say a few words is due to Lorentz.

Matter is entirely formed of electrons bearing enormous charges, and if it appears to us neutral, it is because the electrons' charges of opposite sign balance. For instance, we can picture a kind of solar system consisting of one great positive electron, about which gravitate numerous small planets which are negative electrons, attracted by the electricity of opposite sign with which the central electron is charged. The negative charges of these planets balance the positive charge of the sun, so that the algebraic sum of all these charges is nil.

All these electrons are immersed in ether. The ether is everywhere identical with itself, and perturbations are produced in it, following the same laws as light or the Hertzian oscillations in empty space. Beyond the electrons and the ether there is nothing. When a luminous wave penetrates a part of the ether where the electrons are numerous, these electrons are set in motion under the influence of the perturbation of the ether, and then react upon the ether. This accounts for refraction, dispersion, double refraction, and absorption. In the same way, if an electron was set in motion for any reason, it would disturb the ether about it and give birth to luminous waves, and this explains the emission of light by incandescent bodies.

In certain bodies—metals, for instance—we have motionless electrons, about which circulate movable electrons, enjoying complete liberty, except of leaving the metallic body and crossing the surface that separates it from exterior space, or from the air, or from any other non-metallic body. These movable electrons behave then inside the metallic body as do the molecules of a gas, according to the kinetic theory of gases, inside the vessel in which the gas is contained. But under the influence of a difference of potential the negative movable electrons would all tend to go to one side and the positive movable electrons to the other. This is what produces electric currents, *and it is for this reason that such bodies act as conductors.* Moreover, the velocities of our electrons will become greater as the temperature rises, if we accept the analogy of the kinetic theory of gases. When one of these movable electrons meets the surface of the metallic body, a surface it cannot cross, it is deflected like a billiard ball that has touched the cushion, and its velocity undergoes a sudden change of direction. But when an electron changes its direction, as we shall see further on, it becomes the source of a luminous wave, and it is for this reason that hot metals are incandescent.

In other bodies, such as dielectric and transparent bodies, the movable electrons enjoy much less liberty. They remain, as it were, attached to fixed electrons which attract them. The further they stray, the greater becomes the attraction that tends to bring them back. Accordingly they can only suffer slight displacements; they cannot circulate throughout the body, but only oscillate about their mean position. It is for this reason that these bodies are non-conductors; they are, moreover, generally transparent, and they are refractive because the luminous vibrations are communicated to the movable electrons which are susceptible of oscillation, and a refraction of the original beam of light results.

I cannot here give the details of the calculations. I will content myself with saying that this theory accounts for all the known facts, and has enabled us to foresee new ones, such as Zeeman's phenomenon.

V.

MECHANICAL CONSEQUENCES.

Now we can form two hypotheses in explanation of the above facts.

1. The positive electrons possess an actual mass, much greater than their fictitious electro-magnetic mass, and the negative electrons alone are devoid of actual mass. We may even suppose that, besides the electrons of both signs, there are neutral atoms which have no other mass than their actual mass. In this case Mechanics is not affected, we have no need to touch its laws, actual mass is constant, only the movements are disturbed by the effects of self-induction, as has always been known. These perturbations are, moreover, almost negligible, except in the case of the negative electrons which, having no actual mass, are not true matter.

2. But there is another point of view. We may suppose that the neutral atom does not exist, and that the positive electrons are devoid of actual mass just as much as the negative electrons. But if this be so, actual mass disappears, and either the word *mass* will

have no further meaning, or else it must designate the fictitious electro-magnetic mass; in that case mass will no longer be constant, transversal mass will no longer be equal to longitudinal mass, and the principles of Mechanics will be upset.

And first a word by way of explanation. I said that, for the same charge, the *total* mass of a positive electron is much greater than that of a negative electron. Then it is natural to suppose that this difference is explained by the fact that the positive electron has, in addition to its fictitious mass, a considerable actual mass, which would bring us back to the first hypothesis. But we may equally well admit that the actual mass is nil for the one as for the other, but that the fictitious mass of the positive electron is much greater, because this electron is much smaller. I say advisedly, much smaller. And indeed, in this hypothesis, inertia is of exclusively electro-magnetic origin, and is reduced to the inertia of the ether; the electrons are no longer anything in themselves, they are only holes in the ether, around which the ether is agitated; the smaller these holes are, the more ether there will be, and the greater, consequently, will be its inertia.

How are we to decide between these two hypotheses? By working upon the canal-rays, as Kaufmann has done upon the β rays? This is impossible, for the velocity of these rays is much too low. So each must decide according to his temperament, the conservatives taking one side and the lovers of novelty the other. But perhaps, to gain a complete understanding of the innovators' arguments, we must turn to other considerations.

II.

MECHANICS AND OPTICS.

I.

ABERRATION.

We know the nature of the phenomenon of aberration discovered by Bradley. The light emanating from a star takes a certain time to traverse the telescope. During this time the telescope is displaced by the Earth's motion. If, therefore, the telescope were pointed in the *true* direction of the star, the image would be formed at the point occupied by the crossed threads of the reticule when the light reached the object-glass. When the light reached the plane of the reticule the crossed threads would no longer be in the same spot, owing to the Earth's motion. We are therefore obliged to alter the direction of the telescope to bring the image back to the crossed threads. It follows that the astronomer will not point his telescope exactly in the direction of the absolute velocity of the light from the star—that is to say, upon the true position of the star—but in the direction of the relative velocity of the light in relation to the Earth—that is to say, upon what is called the apparent position of the star.

The velocity of light is known, and accordingly we

might imagine that we have the means of calculating the *absolute* velocity of the Earth. (I shall explain the meaning of this word "absolute" later.) But it is not so at all. We certainly know the apparent position of the star we are observing, but we do not know its true position. We know the velocity of light only in terms of magnitude and not of direction.

If, therefore, the Earth's velocity were rectilineal and uniform, we should never have suspected the phenomenon of aberration. But it is variable: it is composed of two parts—the velocity of the Solar System, which is, as far as we know, rectilineal and uniform; and the velocity of the Earth in relation to the Sun, which is variable. If the velocity of the Solar System—that is to say the constant part—alone existed, the observed direction would be invariable. The position we should thus observe is called the *mean* apparent position of the star.

Now if we take into account at once both parts of the Earth's velocity, we shall get the actual apparent position, which describes a small ellipse about the mean apparent position, and it is this ellipse that is observed.

Neglecting very small quantities, we shall see that the dimensions of this ellipse depend only upon the relation between the Earth's velocity in relation to the Sun and the velocity of light, so that the *relative* velocity of the Earth in relation to the Sun is alone in question.

We must pause, however. This result is not exact, but only approximate. Let us push the approximation a step further. The dimensions of the ellipse will then depend upon the absolute velocity of the Earth.

If we compare the great axes of ellipse for the different stars, we shall have, theoretically at least, the means determining this absolute velocity.

This is perhaps less startling than it seems at first It is not a question, indeed, of the velocity in relation to absolute space, but of the velocity in relation to the ethics, which is regarded, *by definition*, as being in absolute repose.

Moreover, this method is purely theoretical. In fact the aberration is very small, and the possible variations of the ellipse of aberration are much smaller still, and, acccordingly, if we regard the aberration as of the first order, the variations must be regarded as of the second order, about a thousandth of a second of arc, and absolutely inappreciable by our instruments. Lastly, we shall see further on why the foregoing theory must be rejected, and why we could not determine this absolute velocity even though our instruments were ten thousand times as accurate.

Another method may be devised, and, indeed, has been devised. The velocity of light is not the same in the water as in the air: could we not compare the two apparent positions of a star seen through a telescope filled first with air and then with water? The results have been negative; the apparent laws of reflection and of refraction are not altered by the Earth's motion. This phenomenon admits of two explanations.

1. We may suppose that the ether is not in repose, but that it is displaced by bodies in motion. It would not then be astonishing that the phenomenon of refraction should not be altered by the Earth's motion, since everything—lenses, telescopes, and ether—would be carried along together by the same motion. As for

aberration itself, it would be explained by a kind of refraction produced at the surface of separation of the ether in repose in the interstellar spaces and the ether carried along by the Earth's movement. It is upon this hypothesis (the total translation of the ether) that *Hertz's theory* of the Electro-dynamics of bodies in motion is founded.

2. Fresnel, on the contrary, supposes that the ether is in absolute repose in space, and almost in absolute repose in the air, whatever be the velocity of that air, and that it is partially displaced by refringent mediums. Lorentz has given this theory a more satisfactory form. In his view the ether is in repose and the electrons alone are in motion. In space, where the ether alone comes into play, and in the air, where it comes almost alone into play, the displacement is nil or almost nil. In refringent mediums, where the perturbation is produced both by the vibrations of the ether and by those of the electrons set in motion by the agitation of the ether, the undulations are *partially* carried along.

To help us to decide between these two hypotheses, we have the experiment of Fizeau, who compared, by measurements of fringes of interference, the velocity of light in the air in repose and in motion as well as in water in repose and in motion. These experiments have confirmed Fresnel's hypothesis of partial displacement, and they have been repeated with the same result by Michelson. *Hertz's theory, therefore, must be rejected.*

II.

THE PRINCIPLE OF RELATIVITY.

But if the ether is not displaced by the Earth's motion, is it possible by means of optical phenomena to demonstrate the absolute velocity of the Earth, or rather its velocity in relation to the motionless ether? Experience has given a negative reply, and yet the experimental processes have been varied in every possible way. Whatever be the method employed, we shall never succeed in disclosing any but relative velocities; I mean the velocities of certain material bodies in relation to other material bodies. Indeed, when the source of the light and the apparatus for observation are both on the Earth and participate in its motion, the experimental results have always been the same, whatever be the direction of the apparatus in relation to the direction of the Earth's orbital motion. That astronomical aberration takes place is due to the fact that the source, which is a star, is in motion in relation to the observer.

The hypotheses formed up to now account perfectly for this general result, *if we neglect very small quantities on the order of the square of aberration.* The explanation relies on the notion of *local time* introduced by Lorentz, which I will try to make clear. Imagine two observers placed, one at a point A and the other at a point B, wishing to set their watches by means of optical signals. They agree that B shall send a signal to A at a given hour by his watch, and A sets his watch to that hour as soon as he sees the signal. If the operation were performed in this way only, there would be a systematic error; for, since light takes a certain time, *t*, to travel from B to A, A's watch would always be slower than B's to the extent of *t*. This error is easily corrected, for it is sufficient to interchange the signals. A in his turn must send signals to B, and after this new setting it will be B's watch that will be slower than A's to the extent of *t*. Then it will only be necessary to take the arithmetic mean between the two settings.

But this method of operating assumes that light takes the same time to travel from A to B and to return from B to A. This is true if the observers are motionless, but it is no longer true if they are involved in a common transposition, because in that case A, for instance, will be meeting the light that comes from B, while B is retreating from the light that comes from A. Accordingly, if the observers are involved in a common transposition without suspecting it, their setting will be defective; their watches will not show the same time, but each of them will mark the *local time* proper to the place where it is.

The two observers will have no means of detecting this, if the motionless ether can only transmit luminous signals all travelling at the same velocity, and if the other signals they can send are transmitted to them by mediums involved with them in their transposition. The phenomenon each of them observes will be either early or late—it will not occur at the moment it would have if there were no transposition; but since their observations are made with a watch defectively set, they will not detect it, and the appearances will not be altered.

It follows from this that the compensation is easy to

explain so long as we neglect the square of aberration, and for a long time experiments were not sufficiently accurate to make it necessary to take this into account. But one day Michelson thought out a much more delicate process. He introduced rays that had traversed different distances after being reflected by mirrors. Each of the distances being about a yard, and the fringes of interference making it possible to detect differences of a fraction of a millionth of a millimeter ($\frac{1}{25000000}$th of an inch), the square of aberration could no longer be neglected, and yet *the results were still negative.* Accordingly, the theory required to be completed, and this has been done by *the hypothesis of Lorentz and Fitz-Gerald.*

These two physicists assume that all bodies involved in a transposition undergo a contraction in the direction of this transposition, while their dimensions perpendicular to the transposition remain invariable. *This contraction is the same for all bodies.* It is, moreover, very slight, about one part in two hundred million for a velocity such as that of the Earth. Moreover, our measuring instruments could not disclose it, even though they were very much more accurate, since indeed the yard-measures with which we measure undergo the same contraction as the objects to be measured. If a body fits exactly to a measure when the body, and consequently the measure, are turned in the direction of the Earth's motion, it will not cease to fit exactly to the measure when turned in another direction, in spite of the fact that the body and the measure have changed their length in changing their direction, precisely because the change is the same for both. But it is not so if we measure a distance, no longer with a yard-measure, but by the time light takes to traverse it, and this is exactly what Michelson has done.

A body that is spherical when in repose will thus assume the form of a flattened ellipsoid of revolution when it is in motion. But the observer will always believe it to be spherical, because he has himself undergone an analogous deformation, as well as all the objects that serve him as points of reference. On the contrary, the surfaces of the waves of light, which have remained exactly spherical, will appear to him as elongated ellipsoids.

What will happen then? Imagine an observer and a source involved together in the transposition. The wave surfaces emanating from the source will be spheres, having as centre the successive positions of the source. The distance of this centre from the actual position of the source will be proportional to the time elapsed since the emission—that is to say, to the radius of the sphere. All these spheres are accordingly homothetic one to the other, in relation to the actual position S of the source. But for our observer, on account of the contraction, all these spheres will appear as elongated ellipsoids, and all these ellipsoids will still be homothetic in relation to the point S; the excentricity of all the ellipsoids is the same, and depends solely upon the Earth's velocity. *We shall select our law of contraction in such a way that S will be the focus of the meridian section of the ellipsoid.*

This time the compensation is *exact*, and this is explained by Michelson's experiments.

I said above that, according to the ordinary theories,

observations of astronomical aberration could make us acquainted with the absolute velocity of the Earth, if our instruments were a thousand times as accurate, but this conclusion must be modified. It is true that the angles observed would be modified by the effect of this absolute velocity, but the graduated circles we use for measuring the angles would be deformed by the motion; they would become ellipses, the result would be an error in the angle measured, *and this second error would exactly compensate the former.*

This hypothesis of Lorentz and Fitz-Gerald will appear most extraordinary at first sight. All that can be said in its favour for the moment is that it is merely the immediate interpretation of Michelson's experimental result, if we *define* distances by the time taken by light to traverse them.

However that be, it is impossible to escape the impression that the Principle of Relativity is a general law of Nature, and that we shall never succeed, by any imaginable method, in demonstrating any but relative velocities; and by this I mean not merely the velocities of bodies in relation to the ether, but the velocities of bodies in relation to each other. So many different experiments have given similar results that we cannot but feel tempted to attribute to this Principle of Relativity a value comparable, for instance, to that of the Principle of Equivalence. It is well in any case to see what are the consequences to which this point of view would lead, and then to submit these consequences to the test of experiment.

III.

The Principle of Reaction.

Let us see what becomes, under Lorentz's theory, of the principle of the equality of action and reaction. Take an electron, A, which is set in motion by some means. It produces a disturbance in the ether, and after a certain time this disturbance reaches another electron, B, which will be thrown out of its position of equilibrium. Under these conditions there can be no equality between the action and the reaction, at least if we do not consider the ether, but only the electrons *which are alone observable*, since our matter is composed of electrons.

It is indeed the electron A that has disturbed the electron B; but even if the electron B reacts upon A, this reaction, though possibly equal to the action, cannot in any case be simultaneous, since the electron B cannot be set in motion until after a certain length of time necessary for the effect to travel through the ether. If we submit the problem to a more precise calculation, we arrive at the following result. Imagine a Hertz excitator placed at the focus of a parabolic mirror to which it is attached mechanically; this excitator emits electro-magnetic waves, and the mirror drives all these waves in the same direction: the excitator will accordingly radiate energy in a particular direction. Well, calculations show that *the excitator will recoil* like a cannon that has fired a projectile. In the case of the cannon, the recoil is the natural result of the equality of action and reaction. The cannon recoils because the projectile on which it has acted reacts upon it.

But here the case is not the same. What we have fired away is no longer a material projectile; it is energy, and energy has no mass—there is no counterpart. Instead of an excitator, we might have considered simply a lamp with a reflector concentrating its rays in a single direction.

It is true that if the energy emanating from the excitator or the lamp happens to reach a material object, this object will experience a mechanical thrust as if it had been struck by an actual projectile, and this thrust will be equal to the recoil of the excitator or the lamp, if no energy has been lost on the way, and if the object absorbs the energy in its entirety. We should then be tempted to say that there is still compensation between the action and the reaction. But this compensation, even though it is complete, is always late. It never occurs at all if the light, after leaving the source, strays in the interstellar spaces without ever meeting a material body, and it is incomplete if the body it strikes is not perfectly absorbent.

Are these mechanical actions too small to be measured, or are they appreciable by experiment? They are none other than the actions due to the *Maxwell-Bartholi* pressures. Maxwell had predicted these pressures by calculations relating to Electrostatics and Magnetism, and Bartholi had arrived at the same results on Thermodynamic grounds.

It is in this way that *tails of comets* are explained. Small particles are detached from the head of the comet, they are struck by the light of the Sun, which repels them just as would a shower of projectiles coming from the Sun. The mass of these particles is so small that this repulsion overcomes the Newtonian gravitation, and accordingly they form the tail as they retreat from the Sun.

Direct experimental verification of this pressure of radiation was not easy to obtain. The first attempt led to the construction of the *radiometer*. But this apparatus *turns the wrong way*, the reverse of the theoretical direction, and the explanation of its rotation, which has since been discovered, is entirely different. Success has been attained at last by creating a more perfect vacuum on the one hand; and on the other, by not blackening one of the faces of the plates, and by directing a luminous beam upon one of these faces. The radiometric effects and other disturbing causes are eliminated by a series of minute precautions, and a deviation is obtained which is extremely small, but is, it appears, in conformity with the theory.

The same effects of the Maxwell-Bartholi pressure are similarly predicted by Hertz's theory, of which I spoke above, and by that of Lorentz, but there is a difference. Suppose the energy, in the form of light, for instance, travels from a luminous source to any body through a transparent medium. The Maxwell-Bartholi pressure will act not only upon the source at its start and upon the body lighted at its arrival, but also upon the matter of the transparent medium it traverses. At the moment the luminous wave reaches a new portion of this medium, the pressure will drive forward the matter there distributed, and will drive it back again when the wave leaves that portion. So that the recoil of the source has for its counterpart the forward motion of the transparent matter that is in contact with the source; a little later the recoil of this same matter has for its counterpart the forward motion of the transparent matter a little further off, and so on.

Only, is the compensation perfect? Is the action of the Maxwell-Bartholi pressure upon the matter of the transparent medium equal to its reaction upon the source, and that, whatever that matter may be? Or rather, is the action less in proportion as the medium is less refringent and more rarefied, becoming nil in a vacuum? If we admit Hertz's theory, which regards the ether as mechanically attached to matter, so that the ether is completely carried along by matter, we must answer the first and not the second question in the affirmative.

There would then be perfect compensation, such as the principle of the equality of action and reaction demands, even in the least refringent media, even in the air, even in the interplanetary space, where it would be sufficient to imagine a bare remnant of matter, however attenuated. If we admit Lorentz's theory, on the contrary, the compensation, always imperfect, is inappreciable in the air, and becomes nil in space.

But we have seen above that Fizeau's experiment does not permit of our retaining Hertz's theory. We must accordingly adopt Lorentz's theory, and consequently *give up the principle of reaction.*

IV.

Consequences of the Principle of Relativity.

We have seen above the reasons that incline us to regard the Principle of Relativity as a general law of Nature. Let us see what consequences the principle will lead us to if we regard it as definitely proved.

First of all, it compels us to generalize the hypothesis of Lorentz and Fitz-Gerald on the contraction of all bodies in the direction of their transposition. More particularly, we must extend the hypothesis to the electrons themselves. Abraham considered these electrons as spherical and undeformable, but we shall have to admit that the electrons, while spherical when in repose, undergo Lorentz's contraction when they are in motion, and then take the form of flattened ellipsoids.

This deformation of the electrons will have an influence upon their mechanical properties. In fact, I have said that the displacement of these charged electrons is an actual convection current, and that their apparent inertia is due to the self-induction of this current, exclusively so in the case of the negative electrons, but whether exclusively or not in the case of the positive electrons we do not yet know.

On these terms the compensation will be perfect, and in conformity with the requirements of the Principle of Relativity, but only upon two conditions:—

1. That the positive electrons have no real mass, but only a fictitious electro-magnetic mass; or at least

that their real mass, if it exists, is not constant, but varies with the velocity, following the same laws as their fictitious mass.

2. That all forces are of electro-magnetic origin, or at least that they vary with the velocity, following the same laws as forces of electro-magnetic origin.

It is Lorentz again who has made this remarkable synthesis. Let us pause a moment to consider what results from it. In the first place, there is no more matter, since the positive electrons have no longer any real mass, or at least no constant real mass. The actual principles of our Mechanics, based upon the constancy of mass, must accordingly be modified.

Secondly, we must seek an electro-magnetic explanation of all known forces, and especially of gravitation, or at least modify the law of gravitation in the sense that this force must be altered by velocity in the same way as electro-magnetic forces. We shall return to this point.

All this appears somewhat artificial at first sight, and more particularly the deformation of the electrons seems extremely hypothetical. But the matter can be presented differently, so as to avoid taking this hypothesis of deformation as the basis of the argument. Let us imagine the electrons as material points, and enquire how their mass ought to vary as a function of the velocity so as not to violate the Principle of Relativity. Or rather let us further enquire what should be their acceleration under the influence of an electric or magnetic field, so that the principle should not be violated and that we should return to the ordinary laws when we imagine the velocity very low. We shall find that the variations of this mass or of these accelerations must occur *as if* the electron underwent Lorentz's deformation.

V.

Kaufmann's Experiment.

Two theories are thus presented to us: one in which the electrons are undeformable, which is Abraham's; the other, in which they undergo Lorentz's deformation. In either case their mass grows with their velocity, becoming infinite when that velocity becomes equal to that of light; but the law of the variation is not the same. The method employed by Kaufmann to demonstrate the law of variation of the mass would accordingly seem to give us the means of deciding experimentally between the two theories.

Unfortunately his first experiments were not sufficiently accurate for this purpose, so much so that he has thought it necessary to repeat them with more precautions, and measuring the intensity of the fields with greater care. In their new form *they have shown Abraham's theory to be right.* Accordingly, it would seem that the Principle of Relativity has not the exact value we have been tempted to give it, and that we have no longer any reason for supposing that the positive electrons are devoid of real mass like the negative electrons.

Nevertheless, before adopting this conclusion some reflexion is necessary. The question is one of such importance that one would wish to see Kaufmann's experiment repeated by another experimenter.*

* At the moment of going to press we learn that M. Bucherer has repeated the experiment, surrounding it with new precautions, and that, unlike Kaufmann, he has obtained results confirming Lorentz's views.

Unfortunately, the experiment is a very delicate one, and cannot be performed successfully, except by a physicist as skilful as Kaufmann. All suitable precautions have been taken, and one cannot well see what objection can be brought.

There is, nevertheless, one point to which I should wish to call attention, and that is the measurement of the electrostatic field, the measurement upon which everything depends. This field was produced between the two armatures of a condenser, and between these two armatures an extremely perfect vacuum had to be created in order to obtain complete isolation. The difference in the potential of the two armatures was then measured, and the field was obtained by dividing this difference by the distance between the armatures. This assumes that the field is uniform; but is this certain? May it not be that there is a sudden drop in the potential in the neighbourhood of one of the armatures, of the negative armature, for instance? There may be a difference in potential at the point of contact between the metal and the vacuum, and it may be that this difference is not the same on the positive as on the negative side. What leads me to think this is the electric valve effect between mercury and vacuum. It would seem that we must at least take into account the possibility of this occurring, however slight the probability may be.

VI.

The Principle of Inertia.

In the new Dynamics the Principle of Inertia is still true—that is to say, that an *isolated* electron will have a rectilineal and uniform motion. At least it is generally agreed to admit it, though Lindemann has raised objections to the assumption. I do not wish to take sides in the discussion, which I cannot set out here on account of its extremely difficult nature. In any case, the theory would only require slight modifications to escape Lindemann's objections.

We know that a body immersed in a fluid meets with considerable resistance when it is in motion; but that is because our fluids are viscous. In an ideal fluid, absolutely devoid of viscidity, the body would excite behind it a liquid stern-wave, a kind of wake. At the start, it would require a great effort to set it in motion, since it would be necessary to disturb not only the body itself but the liquid of its wake. But once the motion was acquired, it would continue without resistance, since the body, as it advanced, would simply carry with it the disturbance of the liquid, without any increase in the total *vis viva* of the liquid. Everything would take place, therefore, as if its inertia had been increased. An electron advancing through the ether will behave in the same way. About it the ether will be disturbed, but this disturbance will accompany the body in its motion, so that, to an observer moving with the electron, the electric and magnetic fields which accompany the electron would appear invariable, and could only change if the velocity of the electron happened to vary. An effort is therefore required to set the electron in motion, since it is necessary to create the energy of these fields. On the other hand, once the motion is acquired, no effort is necessary to maintain it, since the energy created has only to follow the electron like a wake. This energy, therefore, can only increase the inertia of the electron, as the agitation of the liquid increases that of the body immersed in a perfect fluid. And actually the electrons, at any rate the negative electrons, have no other inertia but this.

In Lorentz's hypothesis, the *vis viva*, which is nothing but the energy of the ether, is not proportional to v^2. No doubt if v is very small, the *vis viva* is apparently proportional to v^2, the amount of momentum apparently proportional to v, and the two masses apparently constant and equal to one another. But *when the velocity approaches the velocity of light, the vis viva, the amount of momentum, and the two masses increase beyond all limit.*

In Abraham's hypothesis the expressions are somewhat more complicated, but what has just been said holds good in its essential features.

Thus the mass, the amount of momentum, and the *vis viva* become infinite when the velocity is equal to that of light. Hence it follows that *no body can, by any possibility, attain a velocity higher than that of light.* And, indeed, as its velocity increases its mass increases, so that its inertia opposes a more and more serious obstacle to any fresh increase in its velocity.

A question then presents itself. Admitting the Principle of Relativity, an observer in motion can have no means of perceiving his own motion. If, therefore, no body in its actual motion can exceed the velocity of light, but can come as near it as we like, it must be the same with regard to its relative motion in relation to our observer. Then we might be tempted to reason as follows:—The observer can attain a velocity of

120,000 miles a second, the body in its relative motion in relation to the observer can attain the same velocity; its absolute velocity will then be 240,000 miles, which is impossible, since this is a figure higher than that of the velocity of light. But this is only an appearance which vanishes when we take into account Lorentz's method of valuing local times.

VII.

The Wave of Acceleration.

When an electron is in motion it produces a disturbance in the ether which surrounds it. If its motion is rectilineal and uniform, this disturbance is reduced to the wake I spoke of in the last section. But it is not so if the motion is in a curve or not uniform. The disturbance may then be regarded as the superposition of two others, to which Langevin has given the names of *wave of velocity* and *wave of acceleration.*

The wave of velocity is nothing else than the wake produced by the uniform motion.

As for the wave of acceleration, it is a disturbance absolutely similar to light waves, which starts from the electron the moment it undergoes an acceleration, and is then transmitted in successive spherical waves with the velocity of light.

Hence it follows that in a rectilineal and uniform motion there is complete conservation of energy, but as soon as there is acceleration there is loss of energy, which is dissipated in the form of light waves and disappears into infinite space through the ether.

Nevertheless, the effects of this wave of acceleration, and more particularly the corresponding loss of energy, are negligible in the majority of cases—that is to say, not only in the ordinary Mechanics and in the motions of the celestial bodies, but even in the case of the radium rays, where the velocity, but not the acceleration, is very great. We may then content ourselves with the application of the laws of Mechanics, stating that the force is equal to the product of the acceleration and the mass, this mass, however, varying with the velocity according to the laws set forth above. The motion is then said to be *quasi-stationary.*

It is not so in all the cases where the acceleration is great, the chief of which are as follows. (1.) In incandescent gases certain electrons take on an oscillatory motion of very high frequency; the displacements are very small, the velocities finite, and the accelerations very great; the energy is then communicated to the ether, and it is for this reason that these gases radiate light of the same periodicity as the oscillations of the electron. (2.) Inversely, when a gas receives light, these same electrons are set in motion with violent accelerations, and they absorb light. (3.) In Hertz's excitator, the electrons which circulate in the metallic mass undergo a sudden acceleration at the moment of the discharge, and then take on an oscillatory motion of high frequency. It follows that a part of the energy is radiated in the form of Hertzian waves. (4.) In an incandescent metal, the electrons enclosed in the metal are animated with great velocities. On arriving at the surface of the metal, which they cannot cross, they are deflected, and so undergo a considerable acceleration, and it is for this reason that the metal emits light.

This I have already explained in Book III., Chap. I., Sec. 4. The details of the laws of the emission of light by dark bodies are perfectly explained by this hypothesis. (5.) Lastly, when the cathode rays strike the anticathode, the negative electrons constituting these rays, which are animated with very great velocities, are suddenly stopped. In consequence of the acceleration they thus undergo, they produce undulations in the ether. This, according to certain physicists, is the origin of the Röntgen rays, which are nothing else than light rays of very short wave length.

III.

THE NEW MECHANICS AND ASTRONOMY.

I.

GRAVITATION.

MASS may be defined in two ways—firstly, as the quotient of the force by the acceleration, the true definition of mass, which is the measure of the body's inertia; and secondly, as the attraction exercised by the body upon a foreign body, by virtue of Newton's law. We have therefore to distinguish between mass, the coëfficient of inertia, and mass, the coëfficient of attraction. According to Newton's law, there is a rigorous proportion between these two coëfficients, but this is only demonstrated in the case of velocities to which the general principles of Dynamics are applicable. Now we have seen that the mass coëfficient of inertia increases with the velocity; must we conclude that the mass coëfficient of attraction increases similarly with the velocity, and remains proportional to the coëfficient of inertia, or rather that the coëfficient of attraction remains constant? This is a question that we have no means of deciding.

On the other hand, if the coëfficient of attraction depends upon the velocity, as the velocities of bodies mutually attracting each other are generally not the same, how can this coëfficient depend upon these two velocities?

Upon this subject we can but form hypotheses, but we are naturally led to enquire which of these hypotheses will be compatible with the Principle of Relativity. There are a great number, but the only one I will mention here is Lorentz's hypothesis, which I will state briefly.

Imagine first of all electrons in repose. Two electrons of similar sign repel one another, and two electrons of opposite sign attract one another. According to the ordinary theory, their mutual actions are proportional to their electric charges. If, therefore, we have four electrons, two positive, A and A′, and two negative, B and B′, and the charges of these four electrons are the same in absolute value, the repulsion of A upon A′ will be, at the same distance, equal to the repulsion of B upon B′, and also equal to the attraction of A upon B′ or of A′ upon B. Then if A and B are very close to each other, as also A′ and B′, and we examine the action of the system A + B upon the system A′ + B′, we shall have two repulsions and two attractions that are exactly compensated, and the resultant action will be nil.

Now material molecules must precisely be regarded as kinds of solar systems in which the electrons circulate, some positive and others negative, *in such a way that the algebraic sum of all the charges is nil.* A material molecule is thus in all points comparable to the system A + B I have just spoken of, so that the total electric action of two molecules upon each other should be nil.

But experience shows us that these molecules attract one another in accordance with Newtonian gravitation, and that being so we can form two hypotheses. We may suppose that gravitation has no connexion with electrostatic attraction, that it is due to an entirely different cause, and that it is merely superimposed upon it; or else we may admit that there is no proportion between the attractions and the charges, and that the attraction exercised by a charge + 1 upon a charge − 1 is greater than the mutual repulsion of two charges + 1 or of two charges − 1.

In other words, the electric field produced by the positive electrons and that produced by the negative electrons are superimposed and remain distinct. The positive electrons are more sensitive to the field produced by the negative electrons than to the field produced by the positive electrons, and contrariwise for the negative electrons. It is clear that this hypothesis somewhat complicates electrostatics, but makes it include gravitation. It was, in the main, Franklin's hypothesis.

Now, what happens if the electrons are in motion? The positive electrons will create a disturbance in the ether, and will give rise in it to an electric field and a magnetic field. The same will be true of the negative electrons. The electrons, whether positive or negative, then receive a mechanical impulse by the action of these different fields. In the ordinary theory, the electro-magnetic field due to the motion of the positive electrons exercises, upon two electrons of opposite sign and of the same absolute charge, actions that are equal and of opposite sign. We may, then, without impropriety make no distinction between the field due to the motion of the positive electrons and the field due to the motion of the negative electrons, and consider only the algebraic sum of these two fields—that is to say, the resultant field.

In the new theory, on the contrary, the action upon the positive electrons of the electro-magnetic field due to the positive electrons takes place in accordance with the ordinary laws, and the same is true of the action upon the negative electrons of the field due to the negative electrons. Let us now consider the action of the field due to the positive electrons upon the negative electrons, or *vice versâ*. It will still follow the same laws, but *with a different coëfficient.* Each electron is more sensitive to the field created by the electrons of opposite denomination than to the field created by the electrons of the same denomination.

Such is Lorentz's hypothesis, which is reduced to Franklin's hypothesis for low velocities. It agrees with Newton's law in the case of these low velocities. More than that, as gravitation is brought down to forces of electro-dynamic origin, Lorentz's general theory will be applicable to it, and consequently the Principle of Relativity will not be violated.

We see that Newton's law is no longer applicable to great velocities, and that it must be modified, for bodies in motion, precisely in the same way as the laws of Electrostatics have to be for electricity in motion.

We know that electro-magnetic disturbances are transmitted with the velocity of light. We shall therefore be tempted to reject the foregoing theory, remembering that gravitation is transmitted, according to Laplace's calculations, at least ten million times as quickly as light, and that consequently it cannot be of electro-magnetic origin. Laplace's result is well known, but its significance is generally lost sight of. Laplace assumed that, if the transmission of gravitation is not instantaneous, its velocity of transmission combines with that of the attracted body, as happens in the case of light in the phenomenon of astronomical aberration, in such a way that the effective force is not directed along the straight line joining the two bodies, but makes a small angle with that straight line. This is quite an individual hypothesis, not very well substantiated, and in any case entirely different from that of Lorentz. Laplace's result proves nothing against Lorentz's theory.

II.

Comparison with Astronomical Observations.

Are the foregoing theories reconcilable with astronomical observations? To begin with, if we adopt them, the energy of the planetary motions will be constantly dissipated by the effect of the *wave of acceleration.* It would follow from this that there would be a constant acceleration of the mean motions of the planets, as if these planets were moving in a resisting medium. But this effect is exceedingly slight, much too slight to be disclosed by the most minute observations. The acceleration of the celestial bodies is relatively small, so that the effects of the wave of acceleration are negligible, and the motion may be regarded as *quasi-stationary.* It is true that the effects of the wave of acceleration are constantly accumulating, but this accumulation itself is so slow that it would certainly require thousands of years of observation before it became perceptible.

Let us therefore make the calculation, taking the motion as quasi-stationary, and that under the three following hypotheses :—

A. Admitting Abraham's hypothesis (undeformable electrons), and retaining Newton's law in its ordinary form.

B. Admitting Lorentz's hypothesis concerning the deformation of the electrons, and retaining Newton's ordinary law.

C. Admitting Lorentz's hypothesis concerning the electrons, and modifying Newton's law, as in the foregoing section, so as to make it compatible with the Principle of Relativity.

It is in the motion of Mercury that the effect will be most perceptible, because it is the planet that has the highest velocity. Tisserand formerly made a similar calculation, admitting Weber's law. I would remind the reader that Weber attempted to explain both the electrostatic and the electro-dynamic phenomena, assuming that the electrons (whose name had not yet been invented) exercise upon each other attractions and repulsions in the direction of the straight line joining them, and depending not only upon their distances, but also upon the first and second derivatives of these distances, that is consequently upon their velocities and their accelerations. This law of Weber's, different as it is from those that tend to gain acceptance to-day, presents none the less a certain analogy with them.

Tisserand found that if the Newtonian attraction took place in conformity with Weber's law, there would result, in the perihelion of Mercury, a secular variation of 14″, *in the same direction as that which has been observed and not explained*, but smaller, since the latter is 38″.

Let us return to the hypotheses A, B, and C, and study first the motion of a planet attracted by a fixed centre. In this case there will be no distinction between hypotheses B and C, since, if the attracting point is fixed, the field it produces is a purely electrostatic field, in which the attraction varies in the inverse ratio of the square of the distance, in conformity with Coulomb's electrostatic law, which is identical with Newton's.

The *vis viva* equation holds good if we accept the new definition of *vis viva*. In the same way the equation of the areas is replaced by another equivalent. The moment of the quantity of motion is a constant, but the quantity of motion must be defined in the new way.

The only observable effect will be a secular motion of the perihelion. For this motion we shall get, with Lorentz's theory, a half, and with Abraham's theory two-fifths, of what was given by Weber's law.

If we now imagine two moving bodies gravitating about their common centre of gravity, the effects are but very slightly different, although the calculations are somewhat more complicated. The motion of Mercury's perihelion will then be 7″ in Lorentz's theory, and 5.6″ in Abraham's.

The effect is, moreover, proportional to n^3a^2, n being the mean motion of the planet, and a the radius of its orbit. Accordingly for the planets, by virtue of Kepler's law, the effect varies in the inverse ratio of $\sqrt{a^5}$, and it is therefore imperceptible except in the case of Mercury.

It is equally imperceptible in the case of the Moon, because, though n is large, a is extremely small. In short, it is five times as small for Venus, and six hundred times as small for the Moon, as it is for Mercury. I would add that as regards Venus and the Earth, the motion of the perihelion (for the same angular velocity of this motion) would be much more difficult to detect by astronomical observations, because the excentricity of their orbits is much slighter than in the case of Mercury.

To sum up, *the only appreciable effect upon astronomical observations would be a motion of Mercury's perihelion, in the same direction as that which has been observed without being explained, but considerably smaller.*

This cannot be regarded as an argument in favour of the new Dynamics, since we still have to seek another explanation of the greater part of the anomaly connected with Mercury; but still less can it be regarded as an argument against it.

III.

Lesage's Theory.

It would be well to set these considerations beside a theory put forward long ago to explain universal gravitation. Imagine the interplanetary spaces full of very tiny corpuscles, travelling in all directions at very high velocities. An isolated body in space will not be affected apparently by the collisions with these corpuscles, since the collisions are distributed equally in all directions. But if two bodies, A and B, are in proximity, the body B will act as a screen, and intercept a portion of the corpuscles, which, but for it, would have struck A. Then the collisions received by A from the side away from B will have no counterpart, or will be only imperfectly compensated, and will drive A towards B.

Such is Lesage's theory, and we will discuss it first from the point of view of ordinary mechanics. To begin with, how must the collisions required by this theory occur? Must it be in accordance with the laws of perfectly elastic bodies, or of bodies devoid of elasticity, or in accordance with some intermediate law? Lesage's corpuscles cannot behave like perfectly elastic bodies, for in that case the effect would be nil, because the corpuscles intercepted by the body B would be replaced by others which would have rebounded from B, and calculation proves that the compensation would be perfect.

The collision must therefore cause a loss of energy to the corpuscles, and this energy should reappear in the form of heat. But what would be the amount of heat so produced? We notice that the attraction passes through the body, and we must accordingly picture the Earth, for instance, not as a complete screen, but as composed of a very large number of extremely small spherical molecules, acting individually as little screens, but allowing Lesage's corpuscles to travel freely between them. Thus, not only is the Earth not a complete screen, but it is not even a strainer, since the unoccupied spaces are much larger than the occupied. To realize this, we must remember that Laplace demonstrated that the attraction, in passing through the Earth, suffers a loss, at the very most, of a ten-millionth part, and his demonstration is perfectly satisfactory. Indeed, if the attraction were absorbed by the bodies it passes through, it would no longer be proportional to their masses; it would be *relatively* weaker for large than for small bodies, since it would have a greater thickness to traverse. The attraction of the Sun for the Earth would therefore be *relatively* weaker than that of the Sun for the Moon, and a very appreciable inequality in the Moon's motion would result. We must therefore conclude, if we adopt Lesage's theory, that the total surface of the spherical molecules of which the Earth is composed is, at the most, the ten-millionth part of the total surface of the Earth.

Darwin proved that Lesage's theory can only lead exactly to Newton's law if we assume the corpuscles to be totally devoid of elasticity. The attraction exercised by the Earth upon a mass 1 at a distance 1 will then be proportional both to S, the total surface of the spherical molecules of which it is composed, to v, the velocity of the corpuscles, and to the square root of ρ, the density of the medium formed by the corpuscles. The heat produced will be proportional to S, to the density ρ, and to the cube of the velocity v.

But we must take account of the resistance experienced by a body moving in such a medium. It cannot move, in fact, without advancing towards certain collisions, and on the other hand retreating before those that come from the opposite direction, so that the compensation realized in a state of repose no longer exists. The calculated resistance is proportional to S, to ρ, and to v. Now we know that the heavenly bodies move as if they met with no resistance, and the precision of the observations enables us to assign a limit to the resistance.

This resistance varying as $S\rho v$, while the attraction varies as $S\sqrt{\rho v}$, we see that the relation of the resistance to the square of the attraction is in inverse ratio of the product Sv.

We get thus an inferior limit for the product Sv. We had already a superior limit for S (by the absorption of the attraction by the bodies it traverses). We thus get an inferior limit for the velocity v, which must be at least equal to 24.10^{17} times the velocity of light.

From this we can deduce ρ and the amount of heat produced. This would suffice to elevate the temperature 10^{26} degrees a second. In any given time the Earth would receive 10^{20} as much heat as the Sun emits in the same time, and I am not speaking of the heat that reaches the Earth from the Sun, but of the heat radiated in all directions. It is clear that the Earth could not long resist such conditions.

We shall be led to results no less fantastic if, in opposition to Darwin's views, we endow Lesage's corpuscles with an elasticity that is imperfect but not nil. It is true that the *vis viva* of the corpuscles will not then be entirely converted into heat, but the attraction produced will equally be less, so that it will only be that portion of the *vis viva* converted into heat that will contribute towards the production of attraction, and so we shall get the same result. A judicious use of the theorem of virial will enable us to realize this.

We may transform Lesage's theory by suppressing the corpuscles and imagining the ether traversed in all directions by luminous waves coming from all points of space. When a material object receives a luminous wave, this wave exercises upon it a mechanical action due to the Maxwell-Bartholi pressure, just as if it had received a blow from a material projectile. The waves in question may accordingly play the part of Lesage's corpuscles. This is admitted, for instance, by M. Tommasina.

This does not get over the difficulties. The velocity of transmission cannot be greater than that of light, and we are thus brought to an inadmissible figure for the resistance of the medium. Moreover, if the light is wholly reflected, the effect is nil, just as in the hypothesis of the perfectly elastic corpuscles. In order to create attraction, the light must be partially absorbed, but in that case heat will be produced. The calculations do not differ essentially from those made in regard to Lesage's ordinary theory, and the result retains the same fantastic character.

On the other hand, attraction is not absorbed, or but very slightly absorbed, by the bodies it traverses, while this is not true of the light we know. Light that would produce Newtonian attraction would require to be very different from ordinary light, and to be, for instance, of very short wave length. This makes no allowance for the fact that, if our eyes were sensible to this light, the whole sky would appear much brighter than the Sun, so that the Sun would be seen to stand out in black, as otherwise it would repel instead of attracting us. For all these reasons, the light that would enable us to explain attraction would require to be much more akin to Röntgen's X rays than to ordinary light.

Furthermore, the X rays will not do. However penetrating they may appear to us, they cannot pass through the whole Earth, and we must accordingly imagine X′ rays much more penetrating than the ordinary X rays. Then a portion of the energy of these X′ rays must be destroyed, as otherwise there would be no attraction. If we do not wish it to be transformed into heat, which would lead to the production of an enormous heat, we must admit that it is radiated in all directions in the form of secondary rays, which we may call X″ rays, which must be much more penetrating even than the X′ rays, failing which they would in their turn disturb the phenomena of attraction.

Such are the complicated hypotheses to which we are led when we seek to make Lesage's theory tenable.

But all that has been said assumes the ordinary laws of Mechanics. Will the case be stronger if we admit the new Dynamics? And in the first place, can we preserve the Principle of Relativity? First let us give Lesage's theory its original form, and imagine space furrowed by material corpuscles. If these corpuscles were perfectly elastic, the laws of their collision would be in conformity with this Principle of Relativity, but we know that in that case their effect would be nil. We must therefore suppose that these corpuscles are not elastic; and then it is difficult to imagine a law of collision compatible with the Principle of Relativity. Besides, we should still get a considerable production of heat, and, notwithstanding that, a very appreciable resistance of the medium.

If we suppress the corpuscles and return to the hypothesis of the Maxwell-Bartholi pressure, the difficulties are no smaller. It is this that tempted Lorentz himself in his Memoire to the Academy of Sciences of Amsterdam of the 25th of April 1900.

Let us consider a system of electrons immersed in an ether traversed in all directions by luminous waves. One of these electrons struck by one of these waves will be set in vibration. Its vibration will be synchronous with that of the light, but there may be a difference of phase, if the electron absorbs a part of the incident energy. If indeed it absorbs energy, it means that it is the vibration of the ether that keeps the electron in vibration, and the electron must accordingly be behind the ether. An electron in motion may be likened to a convection current, therefore every magnetic field, and particularly that due to the luminous disturbance itself, must exercise a mechanical action upon the electron. This action is very slight, and more than that, it changes its sign in the course of the period; nevertheless the mean action is not nil if there is a difference of phase between the vibrations of the electron and those of the ether The mean action is proportional to this difference, and consequently to the energy absorbed by the electron.

I cannot here enter into the details of the calculations. I will merely state that the final result is an attraction between any two electrons varying in the inverse ratio of the square of the distance, and proportional to the energy absorbed by the two electrons.

There cannot, therefore, be attraction without absorption of light, and consequently without production of heat, and it is this that determined Lorentz to abandon this theory, which does not differ fundamentally from the Lesage-Maxwell-Bartholi theory. He would have been still more alarmed if he had pushed the calculations to the end, for he would have found that the Earth's temperature must increase 10^{13} degrees a second.

IV.

Conclusions.

I have attempted to give in a few words as complete an idea as possible of these new doctrines; I have tried to explain how they took birth, as otherwise the reader would have had cause to be alarmed by their boldness. The new theories are not yet demonstrated — they are still far from it, and rest merely upon an aggregation of probabilities sufficiently imposing to forbid our treating them with contempt. Further experiments will no doubt teach us what we must finally think of them. The root of the question is in Kaufmann's experiment and such as may be attempted in verification of it.

In conclusion, may I be permitted to express a wish? Suppose that in a few years from now these theories are subjected to new tests and come out triumphant, our secondary education will then run a great risk. Some teachers will no doubt wish to make room for the new theories. Novelties are so attractive, and it is so hard not to appear sufficiently advanced! At least they will wish to open up prospects to the children, who will be warned, before they are taught the ordinary mechanics, that it has had its day, and that at most it was only good for such an old fogey as Laplace. Then they will never become familiar with the ordinary mechanics.

Is it good to warn them that it is only approximate? Certainly, but not till later on; when they are steeped to the marrow in the old laws, when they have got into the way of thinking in them, and are no longer in danger of unlearning them, then they may safely be shown their limitations.

It is with the ordinary mechanics that they have to live; it is the only kind they will ever have to apply. Whatever be the progress of motoring, our cars will never attain the velocities at which its laws cease to be true. The other is only a luxury, and we must not think of luxury until there is no longer any risk of its being detrimental to what is necessary.

BOOK IV.

ASTRONOMICAL SCIENCE.

I.

THE MILKY WAY AND THE THEORY OF GASES.

THE considerations I wish to develop here have so far attracted but little attention from astronomers. I have merely to quote an ingenious idea of Lord Kelvin's, which has opened to us a new field of research, but still remains to be followed up. Neither have I any original results to make known, and all that I can do is to give an idea of the problems that are presented, but that no one, up to this time, has made it his business to solve.

Every one knows how a great number of modern physicists represent the constitution of gases. Gases are composed of an innumerable multitude of molecules which are animated with great velocities, and cross and re-cross each other in all directions. These molecules probably act at a distance one upon another, but this action decreases very rapidly with the distance, so that their trajectories remain apparently rectilineal, and only cease to be so when two molecules happen to pass sufficiently close to one another, in which case their mutual attraction or repulsion causes them to deviate to right or left. This is what is sometimes called a collision, but we must not understand this word *collision* in its ordinary sense; it is not necessary that the two molecules should come into contact, but only that they should come near enough to each other for their mutual attraction to become perceptible. The laws of the deviation they undergo are the same as if there had been an actual collision.

It seems at first that the orderless collisions of this innumerable dust can only engender an inextricable chaos before which the analyst must retire. But the law of great numbers, that supreme law of chance, comes to our assistance. In face of a semi-disorder we should be forced to despair, but in extreme disorder this statistical law re-establishes a kind of average or mean order in which the mind can find itself again. It is the study of this mean order that constitutes the kinetic theory of gases; it shows us that the velocities of the molecules are equally distributed in all directions, that the amount of these velocities varies for the different molecules, but that this very variation is subject to a law called Maxwell's law. This law teaches us how many molecules there are animated with such and such a velocity. As soon as a gas departs from this law, the mutual collisions of the molecules tend to bring it back promptly, by modifying the amount and direction of their velocities. Physicists have attempted, and not without success, to explain in this manner the experimental properties of gases—for instance, Mariotte's (or Boyle's) law.

Consider now the Milky Way. Here also we see an innumerable dust, only the grains of this dust are no longer atoms but stars; these grains also move with great velocities, they act at a distance one upon another, but this action is so slight at great distances that their trajectories are rectilineal; nevertheless, from time to time, two of them may come near enough together to be deviated from their course, like a comet that passed too close to Jupiter. In a word, in the eyes of a giant, to whom our Suns were what our atoms are to us, the Milky Way would only look like a bubble of gas.

Such was Lord Kelvin's leading idea. What can we draw from this comparison, and to what extent is it accurate? This is what we are going to enquire into together; but before arriving at a definite conclusion, and without wishing to prejudice the question, we anticipate that the kinetic theory of gases will be, for the astronomer, a model which must not be followed blindly, but may afford him useful inspiration. So far celestial mechanics has attacked only the Solar System, or a few systems of double stars. It retired before the aggregations presented by the Milky Way, or clusters of stars, or resoluble nebulæ, because it saw in them only chaos. But the Milky Way is no more complicated than a gas; the statistical methods based upon the calculation of probabilities applicable to the one are also applicable to the other. Above all, it is important to realize the resemblance and also the difference between the two cases.

Lord Kelvin attempted to determine by this means the dimensions of the Milky Way. For this purpose we are reduced to counting the stars visible in our telescopes, but we cannot be sure that, behind the stars we see, there are not others which we do not see; so that what we should measure in this manner would not be the size of the Milky Way, but the scope of our instruments. The new theory will offer us other resources. We know, indeed, the motions of the stars nearest to us, and we can form an idea of the amount and direction of their velocities. If the ideas expounded above are correct, these velocities must follow Maxwell's law, and their mean value will teach us, so to speak, what corresponds with the temperature of our fictitious gas. But this temperature itself depends upon the dimensions of our gaseous bubble. How, in fact, will a gaseous mass, left undisturbed in space, behave, if its elements are attracted in accordance with Newton's law? It will assume a spherical shape; further, in consequence of gravitation, the density will be greater at the centre, and the pressure will also increase from the surface to the centre on account of the weight of the exterior parts attracted towards the centre; lastly, the temperature will increase towards the centre, the temperature and the pressure being connected by what is called the adiabatic law, as is the case in the successive layers of our atmosphere. At the surface itself the pressure will be nil, and the same will be true of the absolute temperature, that is to say, of the velocity of the molecules.

Here a question presents itself. I have spoken of the adiabatic law, but this law is not the same for all gases, since it depends upon the proportion of their two specific heats. For air and similar gases this proportion is 1.41; but is it to air that the Milky Way should be compared? Evidently not. It should be regarded as a monatomic gas, such as mercury vapour, argon, or helium—that is to say, the proportion of the specific heats should be taken as equal to 1.66. And, indeed, one of our molecules would be, for instance, the Solar System; but the planets are very unimportant personages and the Sun alone counts, so that our molecule is clearly monatomic. And even if we take a double star, it is probable that the action of a foreign star that happened to approach would become sufficiently appreciable to deflect the general motion of the system long before it was capable of disturbing the relative orbits of the two components. In a word, the double star would behave like an indivisible atom.

However this may be, the pressure, and consequently the temperature, at the centre of the gaseous sphere are proportional to the size of the sphere, since the pressure is increased by the weight of all the overlying strata. We may suppose that we are about at the centre of the Milky Way, and, by observing the actual mean velocity of the stars, we shall know what corresponds to the central temperature of our gaseous sphere and be able to determine its radius.

We may form an idea of the result by the following considerations. Let us make a simple hypothesis. The Milky Way is spherical, and its masses are distributed homogeneously: it follows that the stars describe ellipses having the same centre. If we suppose that the velocity drops to nothing at the surface, we can calculate this velocity at the centre by the equation of *vis viva*. We thus find that this velocity is proportional to the radius of the sphere and the square root of its density. If the mass of this sphere were that of the Sun, and its radius that of the terrestrial orbit, this velocity, as is easily seen, would be that of the Earth upon its orbit. But in the case we have supposed, the Sun's mass would have to be distributed throughout a sphere with a radius 1,000,000 times as great, this radius being the distance of the nearest stars. The density is accordingly 10^{18} times as small; now the velocities are upon the same scale, and therefore the radius must be 10^{9} as great, or 1,000 times the distance of the nearest stars, which would give about a thousand million stars in the Milky Way.

But you will tell me that these hypotheses are very far removed from reality. Firstly, the Milky Way is not spherical (we shall soon return to this point); and secondly, the kinetic theory of gases is not compatible with the hypothesis of a homogeneous sphere. But if we made an exact calculation in conformity with this theory, though we should no doubt obtain a different result, it would still be of the same order of magnitude: now in such a problem the data are so uncertain that the order of magnitude is the only end we can aim at.

And here a first observation suggests itself. Lord Kelvin's result, which I have just obtained again by an approximate calculation, is in marked accordance with the estimates that observers have succeeded in making with their telescopes, so that we must conclude that we are on the point of piercing the Milky Way. But this enables us to solve another question. There are the stars we see because they shine, but might there not be dark stars travelling in the interstellar spaces, whose existence might long remain unknown? But in that case, what Lord Kelvin's method gives us would be the total number of stars, including the dark stars, and as his figure compares with that given by the telescope, there is not any dark matter, or at least not as much dark as there is brilliant matter.

Before going further we must consider the problem under another aspect. Is the Milky Way, thus constituted, really the image of a gas properly so called? We know that Crookes introduced the notion of a fourth state of matter, in which gases, becoming too rarefied, are no longer true gases, but become what he calls radiant matter. In view of the slightness of its density, is the Milky Way the image of gaseous or of radiant matter? It is the consideration of what is called the *free path* of the molecules that will supply the answer.

A gaseous molecule's trajectory may be regarded as composed of rectilineal segments connected by very small arcs corresponding with the successive collisions. The length of each of these segments is what is called the free path. This length is obviously not the same for all the segments and for all the molecules; but we may take an average, and this is called the *mean free path*, and its length is in inverse proportion to the density of the gas. Matter will be radiant when the mean path is greater than the dimensions of the vessel in which it is enclosed, so that a molecule is likely to traverse the whole vessel in which the gas is enclosed, without experiencing a collision, and it remains gaseous when the contrary is true. It follows that the same fluid may be radiant in a small vessel and gaseous in a large one, and this is perhaps the reason why, in the case of Crookes' tubes, a more perfect vacuum is required for a larger tube.

What, then, is the case of the Milky Way? It is a mass of gas of very low density, but of very great dimensions. Is it likely that a star will traverse it without meeting with any collision—that is to say, without passing near enough to another star to be appreciably diverted from its course? What do we mean by *near enough?* This is necessarily somewhat arbitrary, but let us assume that it is the distance from the Sun to Neptune, which represents a deviation of about ten degrees. Supposing, now, that each of our stars is surrounded by a danger sphere of this radius, will a straight line be able to pass between these spheres? At the mean distance of the stars of the Milky Way, the radius of these spheres will subtend an angle of about a tenth of a second, and we have a thousand million stars. If we place upon the celestial sphere a thousand million little circles with radius of a tenth of a second, will these circles cover the celestial sphere many times over? Far from it. They will only cover a sixteen-thousandth part. Thus the Milky Way is not the image of gaseous matter, but of Crookes' radiant matter. Nevertheless, as there was very little precision in our previous conclusions, we do not require to modify them to any appreciable extent.

But there is another difficulty. The Milky Way is not spherical, and up to now we have reasoned as though it were so, since that is the form of equilibrium that would be assumed by a gas isolated in space. On the other hand, there are clusters of stars whose form is globular, to which what we have said up to this point would apply better. Herschel had already applied himself to the explanation of their remarkable appearance. He assumed that the stars of these clusters are uniformly distributed in such a way that a cluster is a homogeneous sphere. Each star would then describe an ellipse, and all these orbits would be accomplished in the same time, so that at the end of a certain period the cluster would return to its original configuration, and that configuration would be stable. Unfortunately the clusters do not appear homogeneous. We observe a condensation at the centre, and we should still observe it even though the sphere were homogeneous, since it is thicker at the centre, but it would not be so marked. A cluster may, therefore, better be compared to a gas in adiabatic equilibrium which assumes a spherical form, because that is the figure of equilibrium of a gaseous mass.

But, you will say, these clusters are much smaller than the Milky Way, of which it is even probable that they form a part, and although they are denser, they give us rather something analogous to radiant matter. Now, gases only arrive at their adiabatic equilibrium in consequence of innumerable collisions of the molecules. We might perhaps find a method of reconciling these facts. Suppose the stars of the cluster have just sufficient energy for their velocity to become nil when they reach the surface. Then they may traverse the cluster without a collision, but on reaching the surface they turn back and traverse it again. After traversing it a great number of times, they end by being deflected by a collision. Under these conditions we should still have a matter that might be regarded as gaseous. If by chance there were stars in the cluster with greater velocities, they have long since emerged from it, and have left it never to return. For all these reasons it would be interesting to examine the known clusters and try to get an idea of the law of their densities and see if it is the adiabatic law of gases.

But to return to the Milky Way. It is not spherical, and would be more properly represented as a flattened disc. It is clear, then, that a mass starting without velocity from the surface will arrive at the centre with varying velocities, according as it has started from the surface in the neighbourhood of the middle of the disc or from the edge of the disc. In the latter case the velocity will be considerably greater.

Now up to the present we have assumed that the individual velocities of the stars, the velocities we observe, must be comparable to those that would be attained by such masses. This involves a certain difficulty. I have given above a value for the dimensions of the Milky Way, and I deduced it from the observed individual velocities, which are of the same order of magnitude as that of the Earth upon its orbit; but what is the dimension I have thus measured? Is it the thickness or the radius of the disc? It is, no doubt, something between the two, but in that case what can be said of the thickness itself, or of the radius of the disc? Data for making the calculation are wanting, and I content myself with foreshadowing the possibility of basing at least an approximate estimate upon a profound study of the individual motions.

Now, we find ourselves confronted by two hypotheses. Either the stars of the Milky Way are animated with velocities which are in the main parallel with the Galactic plane, but otherwise distributed uniformly in all directions parallel with this plane. If so, observation of the individual motions should reveal a preponderance of components parallel with the Milky Way. This remains to be ascertained, for I do not know that any systematic study has been made from this point of view. On the other hand, such an equilibrium could only be provisional, for, in consequence of collisions, the molecules —I mean the stars—will acquire considerable velocities in a direction perpendicular to the Milky Way, and will end by emerging from its plane, so that the system will tend towards the spherical form, the only figure of equilibrium of an isolated gaseous mass.

Or else the whole system is animated with a common rotation, and it is for this reason that it is flattened, like the Earth, like Jupiter, and like all rotating bodies. Only, as the flattening is considerable, the rotation must be rapid. Rapid, no doubt, but we must understand the meaning of the word. The density of the Milky Way is 10^{25} times as low as the Sun's; a velocity of revolution $\sqrt{10^{25}}$ times smaller than the Sun's would therefore be equivalent in its case from the point of view of the flattening. A velocity 10^{12} times as slow as the Earth's, or the thirtieth of a second of arc in a century, will be a very rapid revolution, almost too rapid for stable equilibrium to be possible.

In this hypothesis, the observable individual motions will appear to us uniformly distributed, and there will be no more preponderance of the components parallel with the Galactic plane. They will teach us nothing with respect to the rotation itself, since we form part of the rotating system. If the spiral nebulæ are other Milky Ways foreign to ours, they are not involved in this rotation, and we might study their individual motions. It is true that they are very remote, for if a nebula has the dimensions of the Milky Way, and if its apparent radius is, for instance, 20″, its distance is 10,000 times the radius of the Milky Way.

But this does not matter, since it is not about the rectilinear motion of our system that we ask them for information, but about its rotation. The fixed stars, by their apparent motion, disclose the diurnal rotation of the Earth, although their distance is immense. Unfortunately, the possible rotation of the Milky Way, rapid as it is, relatively speaking, is very slow from the absolute point of view, and, moreover, bearings upon nebulæ cannot be very exact. It would accordingly require thousands of years of observation to learn anything.

However it be, in this second hypothesis, the figure of the Milky Way would be a figure of ultimate equilibrium.

I will not discuss the relative value of these two hypotheses at any greater length, because there is a third which is perhaps more probable. We know that among the irresoluble nebulæ several families can be distinguished, the irregular nebulæ such as that in Orion, the planetary and annular nebulæ, and the spiral nebulæ. The spectra of the first two families have been determined, and prove to be discontinuous. These nebulæ are accordingly not composed of stars. Moreover, their distribution in the sky appears to depend upon the Milky Way, whether they show a tendency to be removed from it, or on the contrary to approach it, and therefore they form part of the system. On the contrary, the spiral nebulæ are generally considered as independent of the Milky Way: it is assumed that they are, like it, composed of a multitude of stars; that they are, in a word, other Milky Ways very remote from ours. The work recently done by Stratonoff tends to make us look upon the Milky Way itself as a spiral nebula, and this is the third hypothesis of which I wished to speak.

How are we to explain the very singular appearances presented by the spiral nebulæ, which are too regular and too constant to be due to chance? To begin with, it is sufficient to cast one's eyes upon one of these figures to see that the mass is in rotation, and we can even see the direction of the rotation: all the spiral radii are curved in the same direction, and it is evident that it is the *advancing wing* hanging back upon the *pivot*, and that determines the direction of the rotation. But that is not all. It is clear that these nebulæ cannot be likened to a gas in repose, nor even to a gas in relative equilibrium under the domination of a uniform rotation; they must be compared to a gas in permanent motion in which internal currents rule.

Suppose, for example, that the rotation of the central nucleus is rapid (you know what I mean by this word), too rapid for stable equilibrium. Then at the equator the centrifugal force will prevail over the attraction, and the stars will tend to escape from the equator, and will form divergent currents. But as they recede, since their momentum ot rotation remains constant and the radius vector increases, their angular velocity will diminish, and it is for this reason that the advancing wing appears to hang back.

Under this aspect of the case there would not be a true permanent motion, for the central nucleus would constantly lose matter which would go out never to return, and would be gradually exhausted. But we may modify the hypothesis. As it recedes, the star loses its velocity and finally stops. At that moment the attraction takes possession of it again and brings it back towards the nucleus, and accordingly there will be centripetal currents. We must assume that the centripetal currents are in the first rank and the centrifugal currents in the second rank, if we take as a comparison a company in battle executing a turning movement. Indeed the centrifugal force must be compensated by the attraction exercised by the central layers of the swarm upon the exterior layers.

Moreover, at the end of a certain length of time, a permanent status is established. As the swarm becomes curved, the attraction exercised by the advancing wing upon the pivot tends to retard the pivot, and that of the pivot upon the advancing wing tends to accelerate the advance of this wing, whose retrograde motion increases no further, so that finally all the radii end by revolving at a uniform velocity. We may nevertheless assume that the rotation of the nucleus is more rapid than that of the radii.

One question remains. Why do these centripetal and centrifugal swarms tend to concentrate into radii instead of being dispersed more or less throughout, and why are these radii regularly distributed? The reason for the concentration of the swarms is the attraction exercised by the swarms already existing upon the stars that emerge from the nucleus in their neighbourhood. As soon as an inequality is produced, it tends to be accentuated by this cause.

Why are the radii regularly distributed? This is a more delicate matter. Suppose there is no rotation, and that all the stars are in two rectangular planes in such a way that their distribution is symmetrical in relation to the two planes. By symmetry, there would be no reason for their emerging from the planes nor for the symmetry to be altered. This configuration would accordingly give equilibrium, but *it would be an unstable equilibrium.*

If there is rotation on the contrary, we shall get an analogous configuration of equilibrium with four curved radii, equal to one another, and intersecting at an angle of 90°, and if the rotation is sufficiently rapid, this equilibrium may be stable.

I am not in a position to speak more precisely. It is enough for me to foreshadow the possibility that these spiral forms may, perhaps, some day be explained by the help only of the law of gravitation and statistical considerations, recalling those of the theory of gases.

What I have just said about internal currents shows that there might be some interest in a systematic study of the aggregate of the individual motions. This might be undertaken a hundred years hence, when the second edition of the astrographic chart of the heavens is brought out and compared with the first, the one that is being prepared at present.

But I should wish, in conclusion, to call your attention to the question of the age of the Milky Way and the nebulæ. We might form an idea of this age if we obtained confirmation of what we have imagined to be the case. This kind of statistical equilibrium of which gases supply the model, cannot be established except as a consequence of a great number of collisions. If these collisions are rare, it can only be produced after a very long time. If actually the Milky Way (or at least the clusters that form part of it), and if the nebulæ have obtained this equilibrium, it is because they are very ancient, and we shall get an inferior limit for their age. We shall likewise obtain a superior limit, for this equilibrium is not ultimate and cannot last for ever. Our spiral nebulæ would be comparable to gases animated with permanent motions. But gases in motion are viscous and their velocities are finally expended. What corresponds in this case to viscidity (and depends upon the chances of collision of the molecules) is exceedingly slight, so that the actual status may continue for a very long time, but not for ever, so that our Milky Ways cannot be everlasting nor become infinitely ancient.

But this is not all. Consider our atmosphere. At the surface an infinitely low temperature must prevail, and the velocity of the molecules is in the neighbourhood of zero. But this applies only to the mean velocity. In consequence of collisions, one of these molecules may acquire (rarely, it is true) an enormous velocity, and then it will leave the atmosphere, and once it has left it, it will never return. Accordingly our atmosphere is being exhausted exceedingly slowly. By the same mechanism the Milky Way will also lose a star from time to time, and this likewise limits its duration.

Well, it is certain that if we calculate the age of the Milky Way by this method, we shall arrive at enormous figures. But here a difficulty presents itself. Certain physicists, basing their calculations on other considerations, estimate that Suns can have but an ephemeral existence of about fifty millions of years, while our minimum would be much greater than that. Must we believe that the evolution of the Milky Way began while matter was still dark? But how have all the stars that compose it arrived at the same time at the adult period, a period which lasts for so short a time? Or do they all reach it successively, and are those that we see only a small minority as compared with those that are extinct or will become luminous some day? But how can we reconcile this with what has been said above about the absence of dark matter in any considerable proportion? Must we abandon one of the two hypotheses, and, if so, which? I content myself with noting the difficulty, without pretending to solve it, and so I end with a great mark of interrogation. Still, it is interesting to state problems even though their solution seems very remote.

II.

FRENCH GEODESY.*

EVERY one understands what an interest we have in knowing the shape and the dimensions of our globe, but some people would perhaps be astonished at the precision that is sought for. Is this a useless luxury? What is the use of the efforts geodesists devote to it?

If a Member of Parliament were asked this question, I imagine he would answer: "I am led to think that Geodesy is one of the most useful of sciences, for it is one of those that cost us most money." I shall attempt to give a somewhat more precise answer.

The great works of art, those of peace as well as those of war, cannot be undertaken without long studies, which save many gropings, miscalculations, and useless expense. These studies cannot be made without a good map. But a map is nothing but a fanciful picture, of no value whatever if we try to construct it without basing it upon a solid framework. As well might we try to make a human body stand upright with the skeleton removed.

Now this framework is obtained by geodetic meas-

* Throughout this chapter the author is speaking of the work of his own countrymen. In the translation such words as "we" and "our" have been avoided, as far as possible; but where they occur, they must be understood to refer to France and not to England.

urements. Therefore without Geodesy we can have no good map, and without a good map no great public works.

These reasons would no doubt be sufficient to justify much expense, but they are reasons calculated to convince practical men. It is not upon these that we should insist here; there are higher and, upon the whole, more important reasons.

We will therefore state the question differently: Can Geodesy make us better acquainted with nature? Does it make us understand its unity and harmony? An isolated fact indeed is but of little worth, and the conquests of science have a value only if they prepare new ones.

Accordingly, if we happened to discover a little hump upon the terrestrial ellipsoid, this discovery would be of no great interest in itself. It would become precious on the contrary if, in seeking for the cause of the hump, we had the hope of penetrating new secrets.

So when Maupertuis and La Condamine in the eighteenth century braved such diverse climates, it was not only for the sake of knowing the shape of our planet, it was a question of the system of the whole World. If the Earth was flattened, Newton was victorious, and with him the doctrine of gravitation and the whole of the modern celestial mechanics. And to-day, a century and a half since the victory of the Newtonians, are we to suppose that Geodesy has nothing more to teach us? We do not know what there is in the interior of the globe. Mine shafts and borings have given us some knowledge of a stratum one or two miles deep—that is to say, the thousandth part of the total mass; but what is there below that?

Of all the extraordinary voyages dreamed of by Jules Verne, it was perhaps the voyage to the centre of the Earth that led us to the most unexplored regions.

But those deep sunk rocks that we cannot reach, exercise at a distance the attraction that acts upon the pendulum and deforms the terrestrial spheroid. Geodesy can therefore weigh them at a distance, so to speak, and give us information about their disposition. It will thus enable us really to see those mysterious regions which Jules Verne showed us only in imagination.

This is not an empty dream. By comparing all the measurements, M. Faye has reached a result well calculated to cause surprise. In the depths beneath the oceans, there are rocks of very great density, while, on the contrary, beneath the continents there seem to be empty spaces.

New observations will perhaps modify these conclusions in their details, but our revered master has, at any rate, shown us in what direction we must push our researches, and what it is that the geodesist can teach the geologist who is curious about the interior constitution of the Earth, and what material he can supply to the thinker who wishes to reflect upon the past and the origin of this planet.

Now why have I headed this chapter *French Geodesy?* It is because, in different countries, this science has assumed, more perhaps than any other, a national character; and it is easy so see the reason for this.

There must certainly be rivalries. Scientific rivalries

are always courteous, or, at least, almost always. In any case they are necessary, because they are always fruitful.

Well, in these enterprises that demand such long efforts and so many collaborators, the individual is effaced, in spite of himself of course. None has the right to say, this is my work. So the rivalry is not between individuals, but between nations. Thus we are led to ask what share France has taken in the work, and I think we have a right to be proud of what she has done.

At the beginning of the eighteenth century there arose long discussions between the Newtonians, who believed the Earth to be flattened as the theory of gravitation demands, and Cassini, who was misled by inaccurate measurements, and believed the globe to be elongated. Direct observation alone could settle the question. It was the French Academy of Sciences that undertook this task, a gigantic one for that period.

While Maupertuis and Clairaut were measuring a degree of longitude within the Arctic circle, Bouguer and La Condamine turned their faces towards the mountains of the Andes, in regions that were then subject to Spain, and to-day form the Republic of Ecuador. Our emissaries were exposed to great fatigues, for journeys then were not so easy as they are to-day.

It is true that the country in which Maupertuis' operations were conducted was not a desert, and it is even said that he enjoyed among the Lapps those soft creature comforts that are unknown to the true Arctic navigator. It was more or less in the neighbourhood of places to which, in our day, comfortable steamers carry, every summer, crowds of tourists and young English ladies. But at that date Cook's Agency did not exist, and Maupertuis honestly thought that he had made a Polar expedition.

Perhaps he was not altogether wrong. Russians and Swedes are to-day making similar measurements at Spitzbergen, in a country where there are real ice-packs. But their resources are far greater, and the difference of date fully compensates for the difference of latitude.

Maupertuis' name has come down to us considerably mauled by the claws of Dr. Akakia, for Maupertuis had the misfortune to displease Voltaire, who was then king of the mind. At first he was extravagantly praised by Voltaire; but the flattery of kings is as much to be dreaded as their disfavour, for it is followed by a terrible day of reckoning. Voltaire himself learnt something of this.

Voltaire called Maupertuis "my kind master of thought," "Marquess of the Arctic Circle," "dear flattener of the world and of Cassini," and even, as supreme flattery, "Sir Isaac Maupertuis"; and he wrote, "There is none but the King of Prussia that I place on a level with you; his sole defect is that he is not a geometrician." But very soon the scene changes; he no longer speaks of deifying him, like the Argonauts of old, or of bringing down the council of the gods from Olympus to contemplate his work, but of shutting him up in a mad-house. He speaks no more of his sublime mind, but of his despotic pride, backed by very little science and much absurdity.

I do not wish to tell the tale of these mock-heroic

conflicts, but I should like to make a few reflections upon two lines of Voltaire's. In his *Discours sur la Modération* (there is no question of moderation in praise or blame), the poet wrote:—

> Vous avez confirmé dans des lieux pleins d'ennui
> Ce que Newton connut sans sortir de chez lui.
>
> (You have confirmed, in dreary far-off lands,
> What Newton knew without e'er leaving home.)

These two lines, which take the place of the hyperbolical praises of earlier date, are most unjust, and without any doubt, Voltaire was too well informed not to realize it.

At that time men valued only the discoveries that can be made without leaving home. To-day it is theory rather that is held in low esteem. But this implies a misconception of the aim of science.

Is nature governed by caprice, or is harmony the reigning influence? That is the question. It is when science reveals this harmony that it becomes beautiful, and for that reason worthy of being cultivated. But whence can this revelation come if not from the accordance of a theory with experience? Our aim then is to find out whether or not this accordance exists. From that moment, these two terms, which must be compared with each other, become one as indispensable as the other. To neglect one for the other would be folly. Isolated, theory is empty and experience blind; and both are useless and of no interest alone.

Maupertuis is therefore entitled to his share of the fame. Certainly it is not equal to that of Newton, who had received the divine spark, or even of his collaborator Clairaut. It is not to be despised, however, because his work was necessary; and if France, after being outstripped by England in the seventeenth century, took such full revenge in the following century, it was not only to the genius of the Clairauts, the d'Alemberts, and the Laplaces that she owed it, but also to the long patience of such men as Maupertuis and La Condamine.

We come now to what may be called the second heroic period of Geodesy. France was torn with internal strife, and the whole of Europe was in arms against her. One would suppose that these tremendous struggles must have absorbed all her energies. Far from that, however, she had still some left for the service of science. The men of that day shrank before no enterprise—they were men of faith.

Delambre and Méchain were commissioned to measure an arc running from Dunkirk to Barcelona. This time there is no journey to Lapland or Peru; the enemy's squadrons would close the roads. But if the expeditions are less distant, the times are so troublous that the obstacles and even the dangers are quite as great.

In France Delambre had to fight against the ill-will of suspicious municipalities. One knows that steeples, which can be seen a long way off, and observed with precision, often serve as signals for geodesists. But in the country Delambre was working through, there were no steeples left. I forget now what proconsul it was who had passed through it and boasted that he had brought down all the steeples that raised their heads arrogantly above the humble dwellings of the common people.

So they erected pyramids of planks covered with white linen to make them more conspicuous. This was taken to mean something quite different. White linen! Who was the foolhardy man who ventured to set up, on our heights so recently liberated, the odious standard of the counter-revolution? The white linen must needs be edged with blue and red stripes.

Méchain, operating in Spain, met with other but no less serious difficulties. The Spanish country folk were hostile. There was no lack of steeples, but was it not sacrilege to take possession of them with instruments that were mysterious and perhaps diabolical? The revolutionaries were the allies of Spain, but they were allies who smelt a little of the stake.

"We are constantly threatened," writes Méchain, "with having our throats cut." Happily, thanks to the exhortations of the priests, and to the pastoral letters from the bishops, the fiery Spaniards contented themselves with threats.

Some years later, Méchain made a second expedition to Spain. He proposed to extend the meridian from Barcelona to the Balearic Isles. This was the first time that an attempt had been made to cross a large arm of the sea by triangulation, by taking observations of signals erected upon some high mountain in a distant island. The enterprise was well conceived and well planned, but it failed nevertheless. The French scientist met with all kinds of difficulties, of which he complains bitterly in his correspondence. "Hell," he writes, perhaps with some exaggeration, "hell, and all the scourges it vomits upon the earth—

storms, war, pestilence, and dark intrigues—are let loose against me!"

The fact is that he found among his collaborators more headstrong arrogance than good-will, and that a thousand incidents delayed his work. The plague was nothing; fear of the plague was much more formidable. All the islands mistrusted the neighbouring islands, and were afraid of receiving the scourge from them. It was only after long weeks that Méchain obtained permission to land, on condition of having all his papers vinegared—such were the antiseptics of those days. Disheartened and ill, he had just applied for his recall, when he died.

It was Arago and Biot who had the honour of taking up the unfinished work and bringing it to a happy conclusion. Thanks to the support of the Spanish Government and the protection of several bishops, and especially of a celebrated brigand chief, the operations progressed rapidly enough. They were happily terminated, and Biot had returned to France, when the storm burst.

It was the moment when the whole of Spain was taking up arms to defend her independence against France. Why was this stranger climbing mountains to make signals? It was evidently to call the French army. Arago only succeeded in escaping from the populace by giving himself up as a prisoner. In his prison his only distraction was reading the account of his own execution in the Spanish newspapers. The newspapers of those days sometimes gave premature news. He had at least the consolation of learning that he had died a courageous and a Christian death.

Prison itself was not safe, and he had to make his

escape and reach Algiers. Thence he sailed for Marseilles on an Algerian ship. This ship was captured by a Spanish privateer, and so Arago was brought back to Spain, and dragged from dungeon to dungeon in the midst of vermin and in the most horrible misery.

If it had only been a question of his subjects and his guests, the Dey would have said nothing. But there were two lions on board, a present the African sovereign was sending to Napoleon. The Dey threatened war.

The vessel and the prisoners were released. The point should have been correctly made, since there was an astronomer on board; but the astronomer was seasick, and the Algerian sailors, who wished to go to Marseilles, put in at Bougie. Thence Arago travelled to Algiers, crossing Kabylia on foot through a thousand dangers. He was detained for a long time in Africa and threatened with penal servitude. At last he was able to return to France. His observations, which he had preserved under his shirt, and more extraordinary still, his instruments, had come through these terrible adventures without damage.

Up to this point, France not only occupied the first place, but she held the field almost alone. In the years that followed she did not remain inactive, and the French ordnance map is a model. Yet the new methods of observation and of calculation came principally from Germany and England. It is only during the last forty years that France has regained her position.

She owes it to a scientific officer, General Perrier, who carried out successfully a truly audacious enterprise, the junction of Spain and Africa. Stations were established upon four peaks on the two shores of the Mediterranean. There were long months of waiting for a calm and clear atmosphere. At last there was seen the slender thread of light that had travelled two hundred miles over the sea, and the operation had succeeded.

To-day still more daring projects have been conceived. From a mountain in the vicinity of Nice signals are to be sent to Corsica, no longer with a view to the determination of geodetic questions, but in order to measure the velocity of light. The distance is only one hundred and twenty-five miles, but the ray of light is to make the return journey, after being reflected from a mirror in Corsica. And it must not go astray on the journey, but must return to the exact spot from which it started.

Latterly the activity of French Geodesy has not slackened. We have no more such astonishing adventures to relate, but the scientific work accomplished is enormous. The territory of France beyond the seas, just as that of the mother country, is being covered with triangles measured with precision.

We have become more and more exacting, and what was admired by our fathers does not satisfy us to-day. But as we seek greater exactness, the difficulties increase considerably. We are surrounded by traps, and have to beware of a thousand unsuspected causes of error. It becomes necessary to make more and more infallible instruments.

Here again France has not allowed herself to be outdone. Her apparatus for the measurement of bases and of angles leaves nothing to be desired, and I would also mention Colonel Defforges' pendulum, which makes it possible to determine gravity with a precision unknown till now.

The future of French Geodesy is now in the hands of the geographical department of the army, which has been directed successively by General Bassot and General Berthaut. This has advantages that can hardly be overestimated. For good geodetic work, scientific aptitude alone is not sufficient. A man must be able to endure long fatigues in all climates. The chief must know how to command the obedience of his collaborators and to enforce it upon his native helpers. These are military qualities, and, moreover, it is known that science has always gone hand in hand with courage in the French army.

I would add that a military organization assures the indispensable unity of action. It would be more difficult to reconcile the pretensions of rival scientists, jealous of their independence and anxious about what they call their honour, who would nevertheless have to operate in concert, though separated by great distances. There arose frequent discussions between geodesists of former times, some of which started echoes that were heard long after. The Academy long rang with the quarrel between Bouguer and La Condamine. I do not mean to say that soldiers are free from passions, but discipline imposes silence upon over-sensitive vanity.

Several foreign governments have appealed to French officers to organize their geodetic departments. This is a proof that the scientific influence of France abroad has not been weakened.

Her hydrographic engineers also supply a famous

contingent to the common work. The chart of her coasts and of her colonies, and the study of tides, offer them a vast field for research. Finally, I would mention the general levelling of France, which is being carried out by M. Lallemand's ingenious and accurate methods.

With such men, we are sure of the future. Work for them to do will not be wanting. The French colonial empire offers them immense tracts imperfectly explored. And that is not all. The International Geodetic Association has recognized the necessity of a new measurement of the arc of Quito, formerly determined by La Condamine. It is the French who have been entrusted with the operation. They had every right, as it was their ancestors who achieved, so to speak, the scientific conquest of the Cordilleras. Moreover, these rights were not contested, and the French Government determined to exercise them.

Captains Maurain and Lacombe made a preliminary survey, and the rapidity with which they accomplished their mission, travelling through difficult countries, and climbing the most precipitous peaks, deserves the highest praise. It excited the admiration of General Alfaro, President of the Republic of Ecuador, who surnamed them *los hombres de hierro*, the men of iron.

The definitive mission started forthwith, under the command of Lieutenant-Colonel (then Commandant) Bourgeois. The results obtained justified the hopes that had been entertained. But the officers met with unexpected difficulties due to the climate. More than once one of them had to remain for several months at an altitude of 13,000 feet, in clouds and snow, without seeing anything of the signals he had to observe, which refused to show themselves. But thanks to their perseverance and courage, the only result was a delay, and an increase in the expenses, and the accuracy of the measurements did not suffer.

GENERAL CONCLUSIONS.

WHAT I have attempted to explain in the foregoing pages is how the scientist is to set about making a selection of the innumerable facts that are offered to his curiosity, since he is compelled to make a selection, if only by the natural infirmity of his mind, though a selection is always a sacrifice. To begin with, I explained it by general considerations, recalling, on the one hand, the nature of the problem to be solved, and on the other, seeking a better understanding of the nature of the human mind, the principal instrument in the solution. Then I explained it by examples, but not an infinity of examples, for I too had to make a selection, and I naturally selected the questions I had studied most carefully. Others would no doubt have made a different selection, but this matters little, for I think they would have reached the same conclusions.

There is a hierarchy of facts. Some are without any positive bearing, and teach us nothing but themselves. The scientist who ascertains them learns nothing but facts, and becomes no better able to foresee new facts. Such facts, it seems, occur but once, and are not destined to be repeated.

There are, on the other hand, facts that give a large

return, each of which teaches us a new law. And since he is obliged to make a selection, it is to these latter facts that the scientist must devote himself.

No doubt this classification is relative, and arises from the frailty of our mind. The facts that give but a small return are the complex facts, upon which a multiplicity of circumstances exercise an appreciable influence—circumstances so numerous and so diverse that we cannot distinguish them all. But I should say, rather, that they are the facts that we consider complex, because the entanglement of these circumstances exceeds the compass of our mind. No doubt a vaster and a keener mind than ours would judge otherwise. But that matters little; it is not this superior mind that we have to use, but our own.

The facts that give a large return are those that we consider simple, whether they are so in reality, because they are only influenced by a small number of well-defined circumstances, or whether they take on an appearance of simplicity, because the multiplicity of circumstances upon which they depend obey the laws of chance, and so arrive at a mutual compensation. This is most frequently the case, and is what compelled us to enquire somewhat closely into the nature of chance. The facts to which the laws of chance apply become accessible to the scientist, who would lose heart in face of the extraordinary complication of the problems to which these laws are not applicable.

We have seen how these considerations apply not only to the physical but also to the mathematical sciences. The method of demonstration is not the same for the physicist as for the mathematician. But their methods of discovery are very similar. In the case of both they consist in rising from the fact to the law, and in seeking the facts that are capable of leading up to a law.

In order to elucidate this point, I have exhibited the mathematician's mind at work, and that under three forms: the mind of the inventive and creative mathematician; the mind of the unconscious geometrician who, in the days of our far-off ancestors or in the hazy years of our infancy, constructed for us our instinctive notion of space; and the mind of the youth in a secondary school for whom the master unfolds the first principles of the science, and seeks to make him understand its fundamental definitions. Throughout we have seen the part played by intuition and the spirit of generalization, without which these three grades of mathematicians, if I may venture so to express myself, would be reduced to equal impotence.

And in demonstration itself logic is not all. The true mathematical reasoning is a real induction, differing in many respects from physical induction, but, like it, proceeding from the particular to the universal. All the efforts that have been made to upset this order, and to reduce mathematical induction to the rules of logic, have ended in failure, but poorly disguised by the use of a language inaccessible to the uninitiated.

The examples I have drawn from the physical sciences have shown us a good variety of instances of facts that give a large return. A single experiment of Kaufmann's upon radium rays revolutionizes at once Mechanics, Optics, and Astronomy. Why is this? It is because, as these sciences developed, we have recognized more clearly the links which unite them, and at last we have perceived a kind of general design of the map of universal science. There are facts common to several sciences, like the common fountain head of streams diverging in all directions, which may be compared to that nodal point of the St. Gothard from which there flow waters that feed four different basins.

Then we can make our selection of facts with more discernment than our predecessors, who regarded these basins as distinct and separated by impassable barriers.

It is always simple facts that we must select, but among these simple facts we should prefer those that are situated in these kinds of nodal points of which I have just spoken.

And when sciences have no direct link, they can still be elucidated mutually by analogy. When the laws that regulate gases were being studied, it was realized that the fact in hand was one that would give a great return, and yet this return was still estimated below its true value, since gases are, from a certain point of view, the image of the Milky Way; and these facts, which seemed to be of interest only to the physicist, will soon open up new horizons to the astronomer, who little expected it.

Lastly, when the geodesist finds that he has to turn his glass a few seconds of arc in order to point it upon a signal that he has erected with much difficulty, it is a very small fact, but it is a fact giving a great return, not only because it reveals the existence of a little hump upon the terrestrial geoid, for the little hump would of itself be of small interest, but because this hump gives him indications as to the distribution of matter in the interior of the globe, and, through that, as to the past of our planet, its future, and the laws of its development.

THE END.

CPSIA information can be obtained
at www.ICGtesting.com
Printed in the USA
BVHW03*0830070718
520569BV00006B/3/P

9 781616 402549